The Avid® Handbook

The Avid® Handbook

AVID SYMPHONY®, AVID MEDIA COMPOSER®,

AND AVID XPRESS™

THIRD EDITION

Steve Bayes

**Focal
Press**

Boston • Oxford • Auckland • Johannesburg • Melbourne • New Delhi

Butterworth–Heinemann supports the efforts of American
Forests and the Global ReLeaf program in its campaign for
the betterment of trees, forests, and our environment.

ISBN 0-240-80404-X

The publisher offers special discounts on bulk orders of this book.
For information, please contact:
Manager of Special Sales
Butterworth–Heinemann
225 Wildwood Avenue
Woburn, MA 01801-2041
Tel: 781-904-2500
Fax: 781-904-2620

For information on all Butterworth–Heinemann publications available,
contact our World Wide Web home page at: http://www.focalpress.com

10 9 8 7 6 5 4 3 2 1

Printed in the United States of America

To Mary Andress, Marjorie Bayes, and Jane
MacFarlane for all the birthdays

Thanks to:

Peter Bos
Joe Doyle
Matthew Feury
Jamie Fowler
Mark Geffen
Elliott Kaplan
Stephen Hullfish
Dave Meichsner
Jim McKenna
Don Nelsen
Paul Pearman
Greg Staten
Tim Vandawalker

Additional Thanks to:

Wes Plate
Jeannie Munro
James McKenna
Julia Miller
Michael Phillips

"It is circumstances and proper timing that give an action its character and make it either good or bad."

—Agesilaus, 444–400 B.C.

"A globe-full of people, and not one is ignorant of the worth of twenty minutes, each minute is a Pearl, let slip, one after the next, into Oblivion's Gulfs."

—Thomas Pynchon

Table of Contents

Preface

The Avid® Handbook is a book about nonlinear editing that can be read in a linear fashion, but it can also be kept handy to refresh your memory before you start a big job. It will give you the confidence you need after taking an introductory Avid course to deal with the more complex situations that occur outside the classroom.

This is not a tips-and-techniques type of book; there are already several of those and they must be updated every year as new keystrokes are added and other keystrokes are eliminated. A series of keystrokes will only help you under very specific circumstances and does not really represent a coherent view of the machine or its capabilities. Many people who are looking for a tips book are striving for a deeper understanding of ways to make things go faster and smoother and give them an edge over their competition. You won't get that by memorizing ever more arcane variations on secret handshakes. You will gain the power of technology as part of the new production *procedures*, which is the way this book is focused.

Other people will want a beginner's book of "How to Turn It On and Start Making Money," but this type of book devalues the flexibility of this top-of-the-line tool. For better or worse, Avid editing systems are complex because they deal with complex, ever-changing, and unpredictable situations. Take a class or find a good teacher and spend the time with an experienced human being to get going and learn the basics. Having a good teacher is clearly worth the money. It's your career—start this part right!

This book is geared more toward video than film for several reasons. First, there are already several comprehensive books about film and the technology. Second, this technology has completely changed the production methods of the video world, and the impact has not yet been completely explored. Completely digital video production is a reality today, but it's many years off for the film community, and as a result, the procedures in video have changed more radically. Still, much of the material in this book applies to both video and film. Even if you see methods that cannot be applied to your film situation, there may be a spark of recognition as you take several of these techniques and combine them or take the spirit of an idea and modify it.

This book is really intended for overworked editors, assistants, or post-producers who find themselves needing to know more than they really have any chance of learning through their limited experience. They are at a stage bordering on intermediate, but are being asked to make predictions and estimates based on procedures they have never seen and functions they have never used. Everyone in the industry should have a healthy respect for the vast amount of things that can go wrong and increasing levels of complexity *in any* production. A complete overview of procedures will show you everyday savings, shortcuts, and strategies. Display *The Avid® Handbook* proudly in your Avid suite and reread it while you are digitizing and rendering!

Talking to the Machine

"The fact would seem to be, if in my situation one may speak of facts, not only that I shall have to speak of things of which I cannot speak, but also, which is even more interesting, but also that I, which is if possible even more interesting, that I shall have to, I forget, no matter."

—SAMUEL BECKETT, *THE UNNAMABLE*

For artists and craftspeople working today in the communications and entertainment industries, the main characteristic that will distinguish this time period from others will be the speed of change. Tools have always evolved, and clever workmen have always adapted their existing tools to the job at hand. Once a tool was refined to a certain point, it may have stayed in that particular state of grace for centuries. No such luck these days.

Coming to grips with new techniques and features is a constant and never-ending process. You now face an everyday choice of whether to exchange the power of the newest and the fastest for the comfort of the familiar.

But maybe it's not really so bad. Maybe there is a thrill to the new and different, a pleasure in finding the best way to work for you. The secret pleasure of a complex system is simplifying the tools to meet your needs, streamlining them so that you are in complete control. You don't need to buy new software for a new job, just get out the manual and find out what is hidden inside this one giant piece of software. It's a treasure hunt to find the extra bit of power that can propel you through a tedious task and finish while your client is still on the phone. Or like a magic trick, you do have a few things up your sleeve, capabilities your clients have never seen before, things you have developed for yourself that can astonish and amaze.

That's the positive side to all this change. You don't need to know everything. There is no way you can know everything. The time you spend memorizing arcane computer jargon and seldom-used techniques is time away from the creative process and time away from friends and family.

This book is meant to be the next step after formal training. I strongly recommend taking a formal beginner's course. I helped write the courseware (along with a large staff of experienced technical writers and teachers), and I used to certify the instructors. No matter how experienced you are, if you do not start with formal training you will have odd gaps in your knowledge. You are forced to guess and improvise when someone has already thought it out for you. For example, after viewing a demo reel filled with beautiful, multi-layered graphics, I asked the editor to send me his sequence on a floppy. "How do you do that?" he asked. An odd gap, but one that is holding him back.

This handbook is meant to fill in those gaps. It provides the extra level of depth that you can't get in a class full of techniques; the type of experience and insight that comes from watching beginners screw up and advanced people fly; the type of experience that takes a long time to sink in. Along with this handbook you should experiment, read the user documentation, take an advanced course, read the Internet, and never stop learning.

By necessity, training for adults must be short and intense. Unfortunately, it is also hard to retain all the information taught in a short course. When I taught trainers how to teach the Media Composer, my mantra was: "Learn what to leave out." In two, three, or five days you can do more damage trying to overload your students. Make them feel confident and competent with the basics. You will not be "an Avid Editor" after a short course, but you will be able to start and complete your first job. Sometimes I feel an evil joy in telling my students that I am always learning something new on the Avid, after nine years. They don't think that's funny.

A very important translation goes on when you take a vague idea in our heads and translate it into pull-down menus. I have found a little trick that allows the creative to cohabitate with the technical side of my brain. The trick stems from the understanding that the computer is only doing what you ask it to do. The computer is interpreting those commands through the filter of the minds of the people who programmed it. If the designers of editing software don't understand professional post-production, then you will have to fight and trick the machine at almost every step beyond the most basic tasks.

When you sit in front of an Avid interface you trust that the Avid designers understand what you want to do and, if they have not written something to accomplish exactly that, at least they haven't held you back with arbitrary restrictions. You have some freedom to improvise, and many of the techniques described in this book have been developed because the system is so flexible. Many of the techniques have been replaced by simpler, more direct functionality that has been added to the software throughout the years. The writing of this handbook has directly affected my design ideas and manifested itself in minimizing complexity without reducing functionality.

Before I became a senior designer, I taught Avid editing around the world for five years. What was really interesting was that I was learning as much as my students. I learned about sitcoms and army VIP videos, promotion departments, and daily magazine news shows. I saw this same piece of software used every day in many completely different ways. I understood very clearly that there was no right way to do most things, and what is a minor problem for some is a catastrophe for others. Features that you have never seen anyone use will regularly save entire productions. I had to throw away a lot of preconceptions and realize that the more I learned the less I knew.

The value for me in teaching Avid in so many different environments is that I experienced different needs and ways of doing things. The purpose of this handbook is to synthesize all of these experiences into an easy format for you to apply immediately. While using the system for the last nine years, I have seen hundreds of situations where just a little advantage is enough to keep a client or make the difference between losing money, breaking even, or making a profit.

This handbook describes procedures that go beyond the basic functions of the system and adapt to changing situations and demands. Knowing that the procedures are available will allow you to plan more efficient strategies. The more you understand the capabilities for organizing and the database power of Avid's methods, the more you will be able to plan for and stay calm during the inevitable chaos.

So what should you focus on to learn and master the Avid editing systems? Although each model has a different market, there are some very important similarities both in the software and the way it is used. This handbook does not focus on the individual menus or buttons, although they must be covered to get any substantial hands-on value. Buttons and menus change and go out of date so fast that standard printed books no longer can really keep up when you get too specific with professional software (so, OK, this is the third edition in three years, but you know what I mean). The ideas and procedures in this book should help you for some time to come because they deal with the entire process and where the Avid fits in.

The *Avid*® *Handbook* was first written during the beta testing of version 7.0/2.0. The shipping version as of this writing is Macintosh versions 8.1 and 7.2 for Media Composer, and Avid Xpress 2.2; and on Windows NT Avid Xpress version 3.1, Media Composer 9.1, and Symphony 2.1. Details on these models and the benefits of the new hardware will be covered later in the book. Many times a technique for Media Composer or Xpress will apply to Symphony or Xpress NT, so they are not specifically mentioned most of the time. You can count on just about anything that does not have to do with platform differences being the same across all the models.

With so many versions (both recent and legacy), two platforms, and three video boards, it is hard to make sweeping statements on features and functionality. For convenience's sake, I have included both NT and Macintosh modifier keys like this: Ctl/Cmd or Alt/Option. Even though there is a Control key on the Mac, it is not used very often, so when you see the Ctl/Cmd mentioned you should assume the Ctl is on the NT since that is the modifier that maps most closely to the Macintosh Command key. Also, since there are crucial differences in software versions, I will often mention "current" or "latest" to place the release somewhere around 1998 (usually 7.1/2.1 on Mac and later). Occasionally I am forced to specify the exact release in the following form: 3.0/9.0/2.0 on NT. Assume that this applies to Avid Xpress, Media Composer, and Symphony in that order. If you are confused, just give it a try anyway, you might be surprised!

Some of the techniques may not work exactly the same way with the version of the software that you are using. Since this handbook focuses on situations that will not change much, the general ideas and procedures will still be valid even if the pull-down menus get moved around. Unless there are giant leaps in technology, these methods or similar ones, with your own personal variations, should still be in practice. By purchasing the latest versions of any model, you will find new shortcuts and toss out old workarounds. That's what we hope, anyway.

There will be times when a particular function does not exist in Avid Xpress that is standard in Media Composer, but for the most part a technique described in this handbook will make sense anyway and may give you some new ideas about how to use Avid Xpress. Even if you are only using your Avid Xpress once a week, it is still worthwhile to learn these techniques and polish your skills to be more proficient. There is very little that deals specifically with NewsCutter and the newsroom because it is a unique environment. There should be some good material in here for the newsroom, but most of these procedures favor highest quality over quickest method when you are forced to make a choice, although both may be mentioned. I hope you are not forced to make that choice too often!

This handbook makes some assumptions about basic understanding of the Macintosh; otherwise this book would take on the proportions of combined Avid, Macintosh, and Windows NT documentation. This handbook is not a replacement for reading the manuals! There will be some complex techniques to help you deal with complex situations. I like to think of them as recipes, the kind of lists of instructions you deal with every day. I have tried to include the easy techniques, too.

I hope this handbook helps you to avoid down time and maximizes your creative time. Get past the bells and whistles to the really important stuff, and have confidence that with better tools you can become a better editor. You're working with the best system in the world for telling stories with pictures and sound. You will never go back to working the old-fashioned way. In fact, you could forget how tedious it used to be, quickly adapt, and take it all for granted.

2

Workflow of a Nonlinear Project

Workflow is where many beginning editors and producers can take advantage of new opportunities and improve their efficiency by looking at the whole process. The main steps to any nonlinear digital project are input, editing, and output. This chapter describes in-depth strategies for these three basic steps and how to adapt to fit them easily into your production methods. You must get a clear picture of how this technology will change your scheduling and budget.

CHANGE YOUR MIND WITHOUT LOSING IT

First let's deal with the most commonly repeated myth of the benefits of this technology: "It's faster." Well, yes, the computer does arithmetic better than you, but it can't decide between green or blue for a title or whether a logo is too small. You still need to make all the decisions. The computer makes it easier and faster to implement those decisions and to see the many subsequent versions for the client's new vice president, but let's face it, humans don't make decisions faster because they are using a computer. In fact, some things take longer using the computer. Once you have a clear idea of where you want to go, the mechanics of getting there are accelerated.

You would have prudently stopped at the third or fourth version of a sequence using more traditional methods just because of the sheer exhaustion of creating them all or the loudly ticking clock of the expensive online suite. Now you are forced to defend your cut for ethereal reasons of timing, sensibility, and art—reasons that can take much longer to justify. It is also easier to lose that argument since "The client is always right," or at least they are paying the bills. The freedom to experiment and explore can be a liability unless you understand how to take control of a situation, fight for what you think is the best, and know when you are just making the scene worse.

Multiple versions mean multiple screenings for various groups of people with conflicting and passionate opinions. Lots of discussion ensues with arm twisting, insults, shameless self-promotion, and occasionally a decision. This is progress, but it's not fast.

Another reason that your job may not go faster is that since all of these changes are supposed to be so easy and painless, you will get them at the last possible moment. Before, there was a considerable penalty for changing your mind at the last minute, and this put pressure on producers to be better prepared before the final stage of the project. It was easier to charge for the overages because everyone understood how hard those changes would be.

Some changes are easy with a nonlinear system. Making Part 3 into Part 2 takes just a few minutes, but the difference between an easy fix and a much more complicated one may not be obvious. Some changes have always been difficult; for instance, changing layer three in the middle of a twenty-layer effect. Those familiar with traditional layering methods will actually be surprised at how easy it is now to make that kind of change. What they aren't prepared for is all the little changes that ripple throughout the entire show, all the new details that add up to extend a project in unexpected ways. In the workflow of a nonlinear project, there are places where the work will go amazingly fast, but occasionally, the technology and the added capabilities will add some requirements to slow you down.

INPUT

Let's look at the first step: digitizing. If you are working with the Ikegami EditCam, you can skip this step. With versions 7.0/2.0, the ABVB-based Media Composer and Avid Xpress are compatible with the EditCam media. You import the shots from the fieldpak and all your media is online; however, even if you are not digitizing, you still need to incorporate pre-production planning and project organization in a way that will influence the shooting and logging.

Logging

Logging is your link between production and post-production, so don't treat it lightly! If you have strategies in place to minimize hand entry and duplication of efforts to input data to the Avid system, you will have instant savings. Logging and analyzing as much of the footage as possible before post-production really begins is one of the best ways to improve the speed and efficiency of digitizing. If you are working on a film-based project, much of the initial logging will be taken care of for you during the film-to-tape transfer process. The transfer house will create a database, such as a FLExfile, that can be imported as a shotlog into the Film Composer, and the digitizing can start as soon as the film is transferred to video. Editors working on video-based projects must spend more time preparing for the digitizing stage.

Although it may seem tedious at this stage to think deeply about logging, you will find that the more thought you put into the creation of an edit session at the beginning of the process, the more efficient you will be for the rest of the job. I will go into a lot of detail about logging in this chapter even though it seems unimportant to workflow compared to the editing. Ignoring the importance of keeping track of tapes and timecodes is a mistake you make only once!

MediaLog, the logging software that ships with every system, creates Avid bins on a Macintosh without any special hardware. You use a deck control cable that comes with MediaLog and connect it from the tape deck to the serial port of the computer, the connection used for an external modem or a printer. If you can control a video deck with the Macintosh while using MediaLog, you can enter start and end points on the fly for digitizing your clips. MediaLog will read the timecode that is on the address track of the tape. You can then bring the floppy disk with the logged project to the Avid suite and copy the bins to the Avid system. The digitizing is automatically controlled by the computer, which asks for each tape as it is needed.

If you have an internal modem, then the modem serial port on a PowerBook or a Duo is not available, even if it is the only serial port. On a PowerBook computer, you need to turn off the internal modem before you can use that serial port, making the PowerBook a better choice over a Duo. A limited range of professional decks can be controlled without extra deck control hardware (a VLAN or a VLX), so make sure the deck you want to use is on the Avid list of supported decks in the user documentation. Avid uses standard Sony protocol, RS-422, but strangely, many popular decks don't comply with this standard.

Older versions of MediaLog are available for DOS and Windows, but the files need to be converted to Avid Log Exchange (ALE) before they can be imported as a shot log and read by the Macintosh-based editing system. This is a good way to work if you are only logging by eye and not trying to control a deck or if your portable computer is a PC. MediaLog now ships on a Windows-compatible version that is equal to the Macintosh version.

The biggest danger with MediaLog, besides entering the numbers wrong, is logging on a version that is different from the version of Avid software you are using for editing. MediaLog 6.x will have problems sending bins back to Media Composer 5.5. Forward is good, backward is problematic. If you must work with mixed versions of the software (a bad idea), make sure that the version of MediaLog is earlier than the version of Avid editing software. If you are having problems getting your bins into the Avid system, you can always display the most important headings in the bin and export from MediaLog as a shot log. Importing is much more forgiving using a text file instead of an unrecognizable version of a bin.

Timecode Breaks

One of the most important points to consider when logging is timecode breaks in the field tape. Videotape shot in the field will have breaks in the timecode whenever the deck is completely stopped, not just paused; however, if the crew is shooting a time-of-day (TOD) timecode, you will have a break in the timecode every time the deck is paused. The TOD timecode generator continues to increment when the deck is paused and, when the tape starts rolling again, you have a timecode that reflects the new time of day. TOD timecode (sometimes called freerun) is especially problematic with a logging program and specifically with older versions of the Avid software. I do not recommend using TOD unless you are shooting an event where there is continuous action, long takes, and multiple cameras will be running (like multicam). TOD is most useful for a music or stage performance, but a disaster if there are a lot of quick shots. The new exception to this rule is a feature in versions 7.x/2.x called "Use control track instead of time-code for preroll" in the new Digitize User Setting. This allows you to preroll over a break in the timecode during the digitizing if there is no break in the control track, but it still means you will need to be extremely careful while logging.

Avid systems generally create a new master clip whenever they digitize across timecode breaks. You can just pop a tape in, set the Digitize Setting to "Digitize across timecode breaks," and walk away. The Avid software senses every time there is a break in the timecode, makes a new master clip, rolls forward to come up to speed for a few seconds, and then starts digitizing again. If there is plenty of time between the field deck starting and the beginning of the action, this works out fine; however, chances are that at least some of the takes will not come up to speed completely before the action begins and the beginning of the action is not digitized.

Make sure to set the preroll for the deck to be as short as possible. This still may not be enough because the Avid keeps rolling for a few seconds after it finishes digitizing a master clip. If you chop off the beginning of the action or you really want to keep the slate, then you need to occasionally modify the master clip and redigitize. Even if the action is not chopped off, there will be very little useful dead space before the action. You need those pauses for trimming, dissolves, and, as a last resort, room tone. Starting the master clip at the very beginning of the action is a bad habit. Don't make the editing decisions now!

Preroll and postroll are the keys to good logging. You must have a minimum preroll of several seconds after the timecode stabilizes and the field deck comes up to speed. If you are working with Betacam SP, then you only need about three seconds, but you should count on five seconds. U-Matic struggles with anything less than four or five seconds. Don't log the beginning of a shot until this amount of time has elapsed after the field deck has started recording or the deck will rock back and forth during digitizing as it tries to find the proper preroll point — those three to five seconds before the logged inpoint. If the Avid

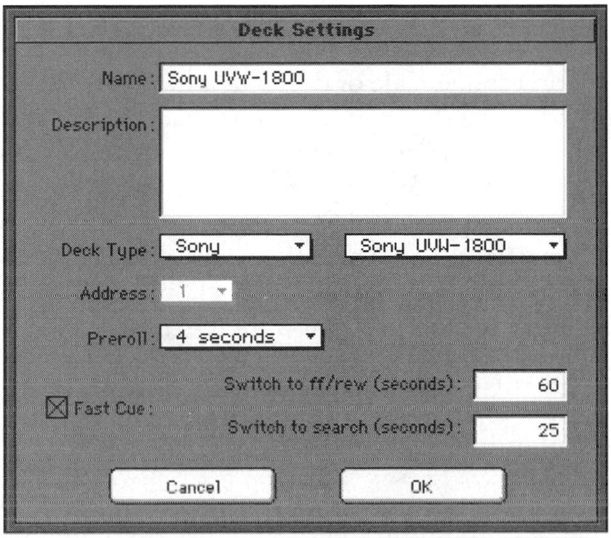

Figure 2.1 Deck Settings Dialog Box

can't find the proper preroll point, it stops and gives you the error message about "failed to find preroll coincidence point." The expected preroll point isn't there.

You can change the deck preroll under the Deck Preferences Setting so that the preroll is shorter. (The default preroll time changes depending on the type of deck you choose in versions 7.x/2.x.) If you set a very short preroll time, you run the risk of the first few frames of the digitized shot being unstable or worse, when you make an edit decision list (EDL), the decks will abort the edit in the online suite. The online editor will have to change the preroll time for that shot, which may stop the auto assembly process, prompting the question: "Who logged this?"

If you absolutely must have the first few frames of a shot and you have a break in the control track, you will not be able to log this shot. You must digitize on the fly by letting the deck roll over the break and quickly clicking on the red digitize button to start the digitizing without preroll; however, if you must make an EDL, you are better off dubbing this shot to another tape because you will never be able to assemble it in the linear online suite.

Using Modify

The best choice you have for fixing a preroll problem is to modify the logged clip and change the inpoint. Highlight the clip and click on the start time in the bin. You can change that number to anything. Alternately, you can go under the Clip menu and pull down Modify. Modify allows you to type in a specific number to increment or decrement the starting timecode of the master clip and allows you to modify all the shots at once in a particular bin. In the case of a preroll problem, either increment the timecode by enough to get the start point away from the control track or timecode break, or decrement the timecode to get a few extra frames from the head of the shot and take your chances on a short preroll.

Modify is a very useful command that will be used many times throughout this book. It was added to the Avid Xpress for version 2.0 but is available on all versions of the Media Composer and Symphony. Be aware that when you modify an inpoint on a master clip, the duration stays the same — the outpoint will move by the same amount. You may have just chopped off the end of the shot! You must also change the outpoint if you want the duration to change when you modify an inpoint.

The Avid does not allow you to change duration to a shot if it has already been completely digitized. This protects you from messing up every potential sequence where this shot was used. Any change to a master clip will ripple across your entire project and affect every sequence and every bin that contains that shot. In order to change duration you must first "undigitize," which is really an "unlink." This allows you to change the inpoint and the duration and unlinks from sequences. This book will discuss linking and unlinking in more detail later.

These are pretty easy fixes compared to the postroll problem. If you log the outpoint of a particular clip after a break in timecode, you will not discover the error until after you have almost finished digitizing the master clip. You may have to wait until the end of a five-minute clip before you get the message that there is a discontinuity in the timecode and the entire material so far must be discarded. Then you must modify the outpoint and determine where the timecode really ends. It pays to double-check any written logs you are handed for these exact reasons. After you type in the handwritten field logs, you should always "go to" both the ins and outs (using the Go To buttons) to check that the numbers exist and have enough preroll and not too much postroll.

There are some methods of logging in the field "on the fly," without stopping the deck, and they may save you some time at the expense of complete accuracy. There are ways to connect the camera or the deck directly to a computer and bang the ins and outs as they happen using infrared signals from the field deck. These methods work best in a controlled environment like a studio, but some people have been known to do it with special battery packs for every piece of equipment.

As you begin to use MediaLog more, you will discover that there are a few shortcuts. For instance, you should use the mouse as little as possible. You can tab through the various entry fields and use the B key or F4 to actually enter the log (make sure the Caps Lock key is off). You can then type the name of the clip and use Control/Command-7 to get back to the logging interface and tab through again, all without using the mouse! Since logging is inherently repetitive, even these small shortcuts will seem like lifesavers later on.

Tape Wear and Dubbing

Some producers are concerned about tape wear and the possibility of creating dropouts or creasing the tapes by playing them too often. This is a serious concern if there is much logging to do. Dropouts occur when small parts of the tape's magnetic coating, the oxide, flake off and stick to the playback heads. This is more noticeable with the smaller formats like Hi-8 or DV since a small amount of oxide on a small format makes a more visible dropout. Although improvements in tape stock are continuing constantly, "prosumer" formats do not find it cost effective to use the more expensive materials or use better quality control when manufacturing these tapes. Usually you will find yourself dubbing to a higher format like Betacam SP and then using those tapes throughout the production while the originals are carefully packed away for a worst-case scenario. If a Betacam SP dub of your Hi-8 is lost or damaged, you have to go back to the original field tapes and re-dub them. If there is timecode on the original tapes, then carry it over to the dubs (called jam-sync). This way you can make another set of dubs to match what you have already digitized. If there is no timecode on the original tapes, you may be forced to re-edit by eye if the dubs are damaged.

In situations where tape wear, a long post-production schedule, and cost are factors, even with a high-end source tape, dub all of your tapes to another format like 3/4-inch U-Matic, S-VHS, or even Betacam SP. Dubbing from a higher format like Digital Betacam to S-VHS or Betacam SP saves you money by reducing the cost of renting an expensive Digital Betacam deck. It also allows you to transfer the timecode so you have a perfect match back to the original sources when you redigitize for the finishing stage (using the original DigiBeta masters and renting the DigiBeta deck for a day or so) or when you make an EDL for the online linear assembly. Evaluate your cost savings and safety factor by the amount of tapes shot for the project and the extra time for dubbing. The cost savings are quickly undone if there are dozens of tapes to dub and dubbing is a cost that must go out of house.

Logging by Hand

To reduce costs and tape wear even further, if you must use the originals when you digitize, there is the choice of dubbing everything to regular VHS just for logging. If you put a timecode window burned into either the upper or lower part of the video frame, you can use this as a visual reference and something you can easily show the client for comments. There is no actual timecode on this tape, just the display of timecode, so you can't actually use these dubs to digitize with. This requires the producer or the editor to view these tapes in the comfort of their own home (or some other place where they don't get charged by the hour) and control the tape with a standard remote control on a VHS deck that allows a good clear freeze frame. Unfortunately, this method requires that someone type by hand the timecode numbers for the start and end of each shot into MediaLog. This option is faster if you purchase the number keypad attachment that connects to your computer's ADB port (where the mouse goes) and have a VHS deck with a search knob for finer control.

Some people have VHS decks that can read VITC, a timecode recorded in the video signal, and can be controlled by the Sony deck protocol, but this is rare. If you have a VHS deck that can read VITC, you can use MediaLog to control the deck and log using the Mark In and Mark Out buttons on the Avid interface. This should reduce pilot error when entering the timecodes. One product, Easy Reader™ from Telcom Research, makes logs based on timecode breaks when controlling a consumer deck if you have recorded VITC on the VHS tape when making the dubs.

There are some challenges with typing numbers in by hand that have to do with reducing human error. First, what kind of timecode has been recorded: drop-frame (DF) or non-drop-frame (NDF)? PAL people don't laugh; someday you will be forced to work in NTSC, too! How can you tell the difference and, um, what is the difference?

Drop-frame vs. Non-Drop-frame

Here is a quick primer on drop-frame and non-drop-frame. NTSC videotape runs at 29.97 frames per second, not 30 frames per second (fps). If you count frames at 30 fps, then you have an interesting problem. A show that uses timecode that increments at 30 fps will actually have a running time longer than one hour even though the timecode says it is exactly 1:00:00:00 for a duration. Because of the slight difference between 29.97 fps (the true playback frame rate) and the more convenient 30 fps, at the end of one hour a non-drop-frame master is too long by exactly 108 frames. This is enough to cut off the credits, or worse, interfere with the commercial coming next! The two frame rates do not change the speed of the tape, just the way that the frames are counted. In fact, both drop-frame and non-drop-frame run at 29.97 fps, but non-drop-frame counts the frames at 30 fps. In order to correct for the frame count not matching playback time, certain frame numbers are skipped in drop-frame timecode. Frames are not actually dropped, but two frame numbers per minute, except for the tenth minute, are not used. This means that when using drop-frame and the timecode says one hour, the real-time playback of the show will take exactly one hour and not one hour, 3 seconds, and 18 frames.

Since accurate timing is most important for broadcast purposes, drop-frame is traditionally used in broadcast. Non-drop-frame is used in corporate work, film transfers for cutting at 24 fps, and other non-broadcast uses like laser discs. This is mostly because the math is easier to do on a standard calculator with non-drop-frame and because all possible frames can be searched by a computer interactive playback system like a kiosk, multi-screen display, or games. There are by hand the timecode numbers for the start and end of each shot into MediaLog. This option is faster if you purchase the number keypad attachment that connects to your computer's ADB port (where the mouse goes) and have a VHS deck with a search knob for finer control.

Some people have VHS decks that can read VITC, a timecode recorded in the video signal, and can be controlled by the Sony deck protocol, but this is rare. If you have a VHS deck that can read VITC, you can use MediaLog to control the deck and log using the Mark In and Mark Out buttons on the Avid interface. This should reduce pilot error when entering the timecodes. One product, Easy Reader™ from Telcom Research, makes logs based on timecode breaks when controlling a consumer deck if you have recorded VITC on the VHS tape when making the dubs.

There are some challenges with typing numbers in by hand that have to do with reducing human error. First, what kind of timecode has been recorded: drop-frame (DF) or non-drop-frame (NDF)? PAL people don't laugh; someday you will be forced to work in NTSC, too! How can you tell the difference and, um, what is the difference?

Drop-frame vs. Non-Drop-frame

Here is a quick primer on drop-frame and non-drop-frame. NTSC videotape runs at 29.97 frames per second, not 30 frames per second (fps). If you count frames at 30 fps, then you have an interesting problem. A show that uses timecode that increments at 30 fps will actually have a running time longer than one hour even though the timecode says it is exactly 1:00:00:00 for a duration. Because of the slight difference between 29.97 fps (the true playback frame rate) and the more convenient 30 fps, at the end of one hour a non-drop-frame master is too long by exactly 108 frames. This is enough to cut off the credits, or worse, interfere with the commercial coming next! The two frame rates do not change the speed of the tape, just the way that the frames are counted. In fact, both drop-frame and non-drop-frame run at 29.97 fps, but non-drop-frame counts the frames at 30 fps. In order to correct for the frame count not matching playback time, certain frame numbers are skipped in drop-frame timecode. Frames are not actually dropped, but two frame numbers per minute, except for the tenth minute, are not used. This means that when using drop-frame and the timecode says one hour, the real-time playback of the show will take exactly one hour and not one hour, 3 seconds, and 18 frames.

Since accurate timing is most important for broadcast purposes, drop-frame is traditionally used in broadcast. Non-drop-frame is used in corporate work, film transfers for cutting at 24 fps, and other non-broadcast uses like laser discs. This is mostly because the math is easier to do on a standard calculator with non-drop-frame and because all possible frames can be searched by a computer interactive playback system like a kiosk, multi-screen display, or games. There are

Tape Wear and Dubbing

Some producers are concerned about tape wear and the possibility of creating dropouts or creasing the tapes by playing them too often. This is a serious concern if there is much logging to do. Dropouts occur when small parts of the tape's magnetic coating, the oxide, flake off and stick to the playback heads. This is more noticeable with the smaller formats like Hi-8 or DV since a small amount of oxide on a small format makes a more visible dropout. Although improvements in tape stock are continuing constantly, "prosumer" formats do not find it cost effective to use the more expensive materials or use better quality control when manufacturing these tapes. Usually you will find yourself dubbing to a higher format like Betacam SP and then using those tapes throughout the production while the originals are carefully packed away for a worst-case scenario. If a Betacam SP dub of your Hi-8 is lost or damaged, you have to go back to the original field tapes and re-dub them. If there is timecode on the original tapes, then carry it over to the dubs (called jam-sync). This way you can make another set of dubs to match what you have already digitized. If there is no timecode on the original tapes, you may be forced to re-edit by eye if the dubs are damaged.

In situations where tape wear, a long post-production schedule, and cost are factors, even with a high-end source tape, dub all of your tapes to another format like 3/4-inch U-Matic, S-VHS, or even Betacam SP. Dubbing from a higher format like Digital Betacam to S-VHS or Betacam SP saves you money by reducing the cost of renting an expensive Digital Betacam deck. It also allows you to transfer the timecode so you have a perfect match back to the original sources when you redigitize for the finishing stage (using the original DigiBeta masters and renting the DigiBeta deck for a day or so) or when you make an EDL for the online linear assembly. Evaluate your cost savings and safety factor by the amount of tapes shot for the project and the extra time for dubbing. The cost savings are quickly undone if there are dozens of tapes to dub and dubbing is a cost that must go out of house.

Logging by Hand

To reduce costs and tape wear even further, if you must use the originals when you digitize, there is the choice of dubbing everything to regular VHS just for logging. If you put a timecode window burned into either the upper or lower part of the video frame, you can use this as a visual reference and something you can easily show the client for comments. There is no actual timecode on this tape, just the display of timecode, so you can't actually use these dubs to digitize with. This requires the producer or the editor to view these tapes in the comfort of their own home (or some other place where they don't get charged by the hour) and control the tape with a standard remote control on a VHS deck that allows a good clear freeze frame. Unfortunately, this method requires that someone type with how to change the type of timecode recorded on their field deck. One free-lance crew may shoot non-drop and the next may shoot drop. This isn't critical, just annoying. To be fair, on some field decks you need to pull out a circuit board to make the change. If you have mixed DF and NDF in your project, you will need to be extremely vigilant during the logging process.

The Importance of Tape Names

When logging in either MediaLog or Avid editing software, start by creating a new tape in the project and give that tape a simple and unique tape name. Tape name is very important at this stage. Short names are good, and a name consist-ing of all numbers is even better, especially if you are planning to create an edit decision list (EDL). When you create a new tape name, you are assigning prop-erties to this tape. Attached to the tape name is an indication whether this tape is drop-frame or non-drop-frame. (This is not an issue in PAL.) The second prop-erty is the name of the project this tape is in. Tape 001 in Project X is a completely different tape than Tape 001 in Project Y. Keep this in mind when you move from a project created in MediaLog to a project created in the Avid editing software.

Keep all tapes in the same project in which they were logged. This means you shouldn't create a new project when you get to the digitizing stage, just copy over the project you started with in MediaLog. If you have logged on software that doesn't make an Avid bin, you will have an ALE (Avid Log Exchange) file that you import to the project. It automatically becomes part of your project, but if you import it twice, the system thinks these are different takes on a different tape with the same name. This will have consequences for your EDL, so don't do it.

There is one more challenge with this method. When your tapes were dubbed, four 30-minute-long field tapes were possibly put onto a 2-hour VHS. When the field tape changes on the dub, treat this as if you had just changed tapes and create a new tape name. When the tape changes, pay attention to what kind of timecode this new tape is. The tape name you created when you logged the first field tape on the VHS will accurately track back to the original field tape.

If you do not change the tape name and log another field tape, you will have problems digitizing. The computer tries to find Tape 001 at timecode 5:15:30:00 when the tape may have changed several times on the VHS dub and the real Tape 001 ended at 1:29:15:00. When digitizing, Tape 001 fast forwards to the end of the reel as the Avid searches for a timecode that is actually on the next tape or several tapes later (probably Tape 005). You will get an error message that the Avid "failed to find timecode coincidence point." You must then tediously find all the shots that start with a five-hour timecode and modify the source name to be Tape 005.

Consistency of timecode type and attention to details that allow you to track back from dubs to the originals is important to prevent confusion. The sys-tem tells you there is a problem when you have logged tape 001 as NDF and you

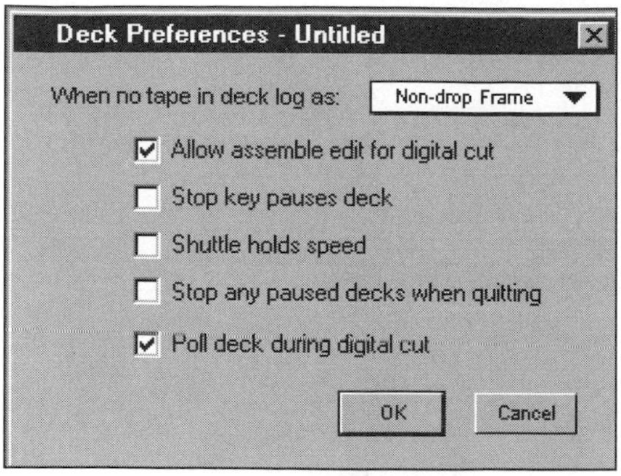

Figure 2.2   Deck Preferences Dialog Box

timecode calculators, however, and the Avid editing systems contain a pretty good one in the Tools menu. Really, once you choose one format or the other, the choice is no longer apparent until it is time to output.

How does the computer know what kind of timecode is on your tapes? In versions 6.5/1.5 and later, when you pop a tape into a deck that is connected to the Avid application, the computer plays a few seconds of that tape automatically. The software reads what kind of code is on the address track before it allows you to name a tape. In earlier versions, you must play the tape yourself for a second in the deck before naming it.

What if you are not controlling the deck with a computer? You must enter the type of timecode by hand. Look at the burn-in timecode in the small window display on the VHS dub. Look at the characters between the last two digits, the frames, and the next two digits, the seconds. If the character is a colon (:), you are working with non-drop-frame. If it is a semicolon (;), you are working with drop-frame. There is a Deck Preferences Setting that says, "If there is no tape in the deck log as" and then a choice between DF and NDF. Well, there is a tape in the deck! But, you are not controlling the deck (it is not physically connected to the computer), so this setting applies. Choose the setting that matches your burn-in timecode and make sure that this type of timecode does not change! If the timecode type in your timecode window dub changes, you have to go back to the Deck Setting and change the type of timecode there.

There is a difference between drop-frame on the field tapes and drop-frame on the finished master. You can use any kind of timecode on the field tapes, and you can mix and match as you create the finished master tape. The master tape will have one type of timecode throughout, and the source tape does not influence this. It is a good idea to have the timecode type in your project be consistent, but sometimes this is not possible because of the use of historical or stock footage. Camera people who always shoot one type of timecode may be unwilling or unfamiliar pop in a DF tape and say "Sure, this is Tape 001" by clicking OK when it asks for Tape 001. Then you will get the following message: "The tape you have entered is drop-frame and has been logged as non-drop-frame or it is non-drop-frame and has been logged as drop-frame." This basically means that the field tape's timecode does not match the tape name that was logged. If this is really the right tape, then a mistake was made during the logging and, when the tape name was assigned, the system was either reading the wrong type of timecode or told to assume the wrong type of timecode because of the Deck Preferences Setting. You must modify clips so they are the right type of timecode and assign them all to a new tape name that has the same type of timecode.

Creating the Database

You are creating the main database for this project whenever you create electronic logs. You can use this database to sort, sift, and search for any and all kinds of information later. Plan to have several custom headings on large jobs that will allow you to search across many bins to find an obscure shot. Although the Avid editing software does not allow you to search across bins, third-party applications will allow you to look inside even closed bins. Retrieve-It! from MVP Solutions is the best program I have seen for this because it allows you to peek inside the programming code of a closed bin without actually opening it. Since large bins take some time to open, this can really allow you to fly along. Keep in mind, though, that a third-party search engine like Retrieve-It! does not know to look for the 15 in the Scene column. Retrieve-It! doesn't know about columns. Be sure that any search criteria you create is valid without the column information, like "15 Lake view WS."

Alternatives to MediaLog

You don't need to use MediaLog to log your project. You can use any word processing, spreadsheet, or database program as long as you format it correctly. There is a standard way to set up the headings and then tab-delimit the information according to the format set forth in the user documentation. You can then convert the file to a text-only document and import to Avid editing software as a shot log. You can create a template in a program like FileMaker Pro and create your bin information to import to the Avid to fit that template.

If you export from Avid editing software as a text file that is tab-delimited, then FileMaker Pro enters the column information into the right headings in your template. You can then search on 15 in the Scene field of the database template. This takes some setting up but can be very powerful.

MediaLog ships with every Avid editing system and is by far the simplest to use since it is almost identical to the digitizing interface of the Avid editing software. Some people like other programs because with MediaLog you can't start entering information about a master clip until you mark an outpoint. You can fast forward to the end of a clip and mark it and then enter all the heading information. But if you want to view the tape in real time and enter the heading information at the same time, then you might want to consider a database or a spreadsheet program. There are also programs on the market, like Executive Producer by Imagine Products, which create complex databases using the Avid text format with screen grabs and multiple ways to format the information for presentation purposes. Executive Producer is a good alternative to MediaLog if you need a more extensive database but want it to be formatted for Avid import.

Organizing Bins

An editing system so heavily dependent on the computer means you must spend more time getting organized at the beginning of the job. It may appear like you are off to a slow start, but again, when you get to the fast part—the editing—you will look like a mind reader. The most successful method for organizing bins is to start with a bin for each field tape and then duplicate clips and move them to content-based bins afterward. This allows you to print a copy of the contents of each tape as it is logged and keep a central notebook of all the tapes. You can go back to the tape months later and know exactly what is on that tape when you archive it. If the worst should happen and you cannot continue the job on the Avid system, then you will have something tangible to take with you to another edit suite.

Of course all of these strategies assume that you have time to prepare! There will be times when the client just walks into the production house with a box of tapes and literally drops them on you. Now you must log, digitize, and view all at the same time. Some people, not wanting to miss anything, will start to digitize by figuring out how much space is available and how much material is there and digitize every tape from beginning to end. There are two nice utilities called StorageCalc (in your Mac Utilities folder) and Storage Tool by Tony Black (aeblack@randomvideo.com). Both of these can help you with the math. Storage Tool looks at the available storage space and does the calculations based on AVR. Digitizing everything is a good strategy if you are not going to be the editor because it leaves nothing out.

There are times when you must make creative decisions on the fly before you have seen all the material, and here you can only operate on your gut instinct and experience. Scan through the tapes as fast as you can and log what appears to be the best and second-best take of every set up. Go back and batch digitize after you have scanned through everything. The benefit to this method is that by the time you get around to actually digitizing, you will have seen everything and can weed out the redundant and superfluous. Be sure to grab some shots that you think are cool and have no idea what to do with—they may be the most useful for tying sections together with graphics.

By far the worst outcome of this scenario is when the client tells you exactly what they want, exactly where it is, and that is all you digitize. Don't be fooled! You will be looking for shots on tape all day, and you will lose almost all of the power of your random access tool. It is imperative that you see everything. They brought the job to you because they trust your judgment, but if you can't see everything then you can't make the creative leap they expect from you under pressure.

Batch Digitizing

When you finally get to the digitizing step, you need to have a standard procedure and follow it closely. You are responsible for technically reinterpreting this footage for everyone else who is going to work on this project. Make sure you can justify every level you adjust. The last thing you want to hear while you are editing furiously is "Does it really look like that on the tape?" There is a button on the Media Composer just for those times, called Find Frame, which asks for the original tape to be mounted and then fast forwards to the exact frame. If you have a roomful of nervous people or a client who is inexperienced with compressed images, you may want to map this button and keep your tapes close by. This is an inevitable slowdown, but if your clients are unsure, then you should take the extra effort to reassure them, especially at the beginning of the project. As they learn to trust you and your judgment, they should request this less (unless there really is something wrong with the originals).

Also, keep in mind that the Avid allows you to save those video levels on a tape-by-tape basis. Those video levels may be the ones used when it comes time to finish. Chapter 10 discusses the setting of levels.

A digital project is never really finished. The mess you leave today will come back to you in a month or two as the clients come back for version 2! Or the next editor will curse you loudly as they try to figure out why you put things in random bins and left all of those unnamed clips and sequences. Mastering the first part of this workflow will impact the rest of the job. It will make the difference between always rushing to catch up as you search frantically or staying one step ahead, concentrating on creativity when it is most required, during the editing.

EDITING FLOW

This is the part you should give thanks for every day, freeing you from the restrictions and limitations you have been struggling against since the first time someone showed you how to load a tape or pull apart a splice. Editing is also the most intimidating step because, without the restrictions, many of the excuses for not being perfect have long disappeared. Some editors fight back by quickly claiming something just can't be done, but be careful, the next editor that works with this director may know how to do it! This is the double edge of the new tool: You are always struggling between what you know how to do fast and confidently and that nagging feeling in the back of your head that if you just had a half an hour you could figure out a better way.

Part of this anxiety comes because we are still trying to use the new tools in the old way. Sometimes we focus on the one best way to do something and bring all of our old skills to bear on the new technology. These skills were honed and perfected because of the demands and limitations of the old tools. This is one of the biggest reasons why experienced editors don't want to spend the time to learn all the tips and tricks. The new tools seem too technical, and they just want to edit.

It is easy to get bogged down in option keys and which direction to drag the mouse, but these details are the last things you should be concentrating on in the beginning. What you should be thinking about is how to take what is truly valuable and brilliant about the way you work now and translate it to the new way of working. In the process, you should take a long, hard look at those skills you have developed only because they were necessary and discard them if they are no longer so. I used to be proud of how fast I could thread a one-inch machine. The one good part about threading a one-inch machine is that it gets you out of the chair, so you can stretch and think for a minute. That is the real importance of the skill and the valuable part to keep!

There is a lot of personal choice involved in discovering the "right" ways to do particular functions on the Avid editing software. The more you work on the system, the quicker you can identify when a technique is not as good as another in a specific situation. Remember the technique or feature that is easiest and most intuitive for you right now. Get one way down and feel confident about that. Later, work on finding the best way, but don't let your fear of not knowing the absolute best way paralyze you and keep you from using the way you know. I used to pretend to feel ill, then grab the manual and read it in the bathroom.

The more complicated techniques are valuable since, at some time in the future, you can count on getting a job that will require them. At the beginning, focus on the content of what you are working on and don't let the machine get in the way. The faster you can reach this stage of transparency with the machine, the better you will feel about the new tools. Only then should you think about memorizing the more advanced techniques.

Refine, Refine, Refine

Do what you do best — edit — and let the machine do what it does best — give you access to your images and allow you to see what is in your head faster. For some editors, if not most, the vision they have comes about through refinement. They have a rough idea of where the scene is to go or the best flow for the material, and they start off that way only to find something better, or something missing. Unless you are working with an incredibly tightly scripted show and very little footage, there is a lot of exploration and discovery as you get more familiar with the material. Just like the speed of changing a one-inch reel becomes less important, the requirement to get a cut absolutely perfect on the first try also becomes less important. The digital nonlinear technology is perfect for this type of approach.

The first pass on a sequence is truly a rough cut. Although we have used this term in the past, it usually meant that you tried your best to get everything perfect and afterward there were some changes. Now it means more precisely that you have roughly assembled the parts and are planning a series of creative refinements.

I am tempted to not show my first pass to anyone. In particular, I don't want to show it to producers who are not familiar with the possibilities of nonlinear. They would think I was the worst editor they had ever worked with and think about firing me on the spot! I watched a young producer squirm as their VP of communication saw the first rough cut. It had no effects or graphics and had burned-in timecode. The producer explained very carefully about what they were about to show, but the VP still blew his top. It took everyone in the room to reassure the VP that this was all perfectly normal.

Your first nonlinear assembly should be even rougher than this. Many times just taking all the good takes and lining them up one after the other should be your first step. I have worked with clients who were always vaguely unsure if they were using the right take, even after going through an exhaustive review. This is a legitimate concern, especially since every shot depends on all the shots around it. Changing shot seven may make you reconsider shot fifteen. In the past, there was a lot of tape or film shuttling, and this was a good way to remind yourself (and everyone watching) what you had to work with. It was also a good excuse to occasionally just look at all the takes one more time. Again, this is one of the valuable old skills that should be brought along to the new tools. Do not discard the time spent watching and thinking because you are not forced to view the material over and over.

Take the circled takes and string them together. Take all the best cutaways and make a true B roll that you can use as a source. Use the match frame command, not to extend your shot as so many beginners do, but to remind yourself what came after your original cut point. Use the time gained by not having to dub or rip apart splices to think, watch, and explore. You may not end up cutting faster, but you will cut better and be happier with the end result. Refine, refine, and refine. Start with big chunks and whittle them down. During the first several passes, work for content, then work on flow and timing (if you can separate them).

Use a strategy that incorporates each step only when it is necessary. You should not be worrying about the graphics when you are still trying to work out the soundbites. Some people call this stage the radio cut because it is based on stringing all the soundbites together to tell the story based on spoken word. Finally, go for the effects and the smoothing of the audio mix. This particular order is totally up to you. If you cut with a full sound mix as your style, then put that as the second pass, but don't try to do everything at once.

You will find yourself watching the scene or segment over and over. In the past, you may not have taken the time to do that until you were almost done. Look at the big picture, pull yourself out of the details until the final stages where they belong.

There is room for reconsideration and the type of visual juxtaposition that you might never have thought of unless it was right there in front of you. The ability to hold a scene in your head is a very valuable skill, somewhat like memorizing the opening 50 moves of a chess game. But that is a mental skill, not a visual one. The instruction I give beginners as I see them staring at the screen is: "Don't guess!" Don't create scenarios in your head when you can create them with your eyes. The speed with which you can put two disparate and unexpected images together is another tool for shaking up the other side of your brain. Work beyond the everyday logic and justifications you would have used before even attempting a sequence. Seeing new things with your eyes and not your head will inspire you.

WORKFLOW OUTSIDE THE SUITE

A really good producer can foresee what parts of a project should be left to an editor at an NLE workstation and what parts other experts should take care of. There are two basic reasons to send parts of the project out of the suite — time and expertise.

WORKING WITH GRAPHICS

If the graphics can be done in tandem with the editing, then, when they are needed at the end of the finishing stage, they can just be inserted. Contrast this with the producer who gets to some midpoint and says, "Let's come up with 'a look' now." All editing stops and you become a graphic designer, a 3-D animator, and an expert on four or five third-party programs. Some editors like this new challenge and the creative control, and more power to them. The editor is probably the best person to understand how all of the effects and graphics should be incorporated into the show and the right formats and work parts necessary to make it happen. But editors must also make the correct estimate as to how long the third-party design and render time is going to take on any particular computer they are working with. A wrong guess or the failure to build the time into the overall schedule will grind everything to a halt while they watch a revolving cursor and decide whether to come back tomorrow.

Sending all the graphics out takes care of the time and expertise problems but replaces them with new ones. What if the graphics are not right? What if they are the wrong color or the animations are not long enough to cover the new voice over? The time involved to send them back to the (unsupervised) graphic artist will surely halt the post on your project. A smart producer needs to be able to monitor the bits and pieces being done out of sight just as much as keeping an eye on the progress of the edit. This planning process must also schedule the time to add the graphics at a stage when they will mesh nicely with the editor's final finishing schedule.

All of these concerns are not new. The problem is that in the past, sending the non-editorial bits out to another artist was the only way to get a highly professional finished product. With the onset of the "everything in a box" concept, producers seem to be ignoring their time-honored knowledge, and all those who are getting in for the first time have no experience with anything except this new desktop method.

SOUND MIXING AND COLOR CORRECTION

Many times people ask about color correcting and sound mixing. Yes, the "box" can do it. Who is running the box? Do you play musical chairs as the sound mixer sits down in your chair and fiddles with the sweetening for a while before they let you back in the room? Not likely. Either you are expected to be a sound engineer and a colorist or the process needs to be segmented into parts.

There are some basic challenges with the workflow of sound mixing. You are dealing with a chicken-and-egg scenario. Which comes first: the sound or the pictures? The answer is "Yes." Thankfully, there is Avid's non-proprietary exchange format OMFI (Open Media Format Interchange). OMFI allows you to send compositions and digitized sound media to digital audio workstations like AudioVision or Pro Tools. This means that there can be parallel processing of sound while the cutting continues. I have held a telephone up to a speaker so someone in New York City could record and use it for a scratch track, but this wasn't something anyone actually planned for. The producer decided that a New York City talent was the only thing that would do and we waited, not so patiently, for the next shuttle to Boston to bring us the master mix. It did help get the governor elected, however — no, not Dukakis!

The give and take of the sound track cannot be underestimated. That extra few piano notes before a transition might change the length of the dissolve or the amount of time in black. You are shortchanging yourself if you just send the project out for a mix and never see it again until after it is finished and dubbed. There should be time built into the schedule to tweak the cut with the new music, but this is a perfect world scenario. If you can get your timings and transitions as close as possible using scratch tracks and click tracks, then you are way ahead of the game. But wouldn't it be great if you could use the time that you are rendering or waiting for final client approval to coincide with the sound being mixed? What a perfect excuse to wait until tomorrow and get a chance for another "final" pass at the finished product with the finished sound track in place.

The alternative to sending the sound out is, of course, DIY, do it yourself. Lately, there have been some great additions to the sound mixing capabilities of the Avid editing systems. Real-time level adjustment with keyframes, equalization, waveforms, and audio punch-in are all features you have come to expect. The newest release features Ride and Remember, which allows you to adjust the faders of an external MIDI control surface. Avid editing software remembers those levels and allows you to further edit them. You end up with lots of keyframes that can then be filtered down to the minimal amount you need. Another new feature is the ability to use Audio Suite plug-ins for all the special effects that you couldn't do before with just EQ-like pitch changing and reversing the sound.

Many people feel that sound mixing will never be complete without subframe editing, the ability to make changes at the sampling level, which gives much more control than what is available at 30, 25, or 24 frames per second. Fortunately, Avid digitized sound media is either in the native SDII format (Sound Designer) or AIFF, which is compatible with other sound mixing programs like Pro Tools or AudioVision. You can export sequence information as OMFI, open existing audio media, and start mixing. This book will discuss audio importing and exporting issues in Chapter 6.

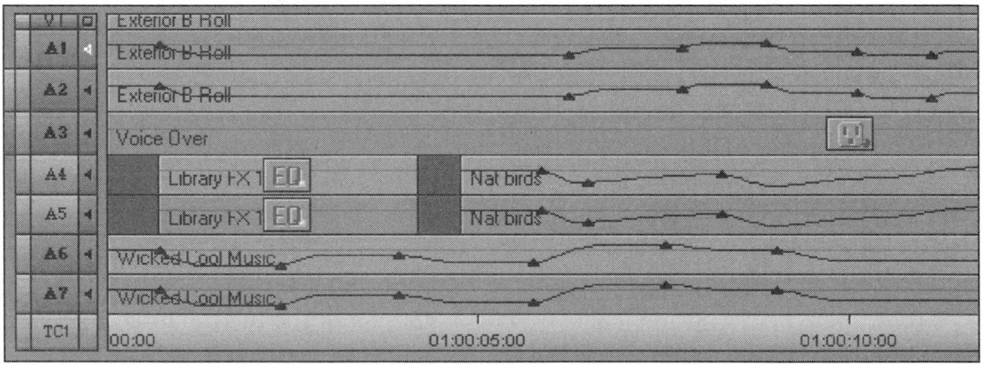

Figure 2.3 Rubberbanding, EQ, and AudioSuite Plug-ins with 8-Channel Monitoring

Color Correction

Color correction has been, up until now, a linear process—to adjust shot two, you need to see shots one and three. In a nonlinear project, shots are digitized in an order that reduces tape shuttling and tape changes, but does nothing to allow the editor to see two juxtaposed shots one after the other until later. There are built-in color correction features that allow a wide range of adjustment in the whole range of Avid editing systems. Those tools have been continuously improved over the latest several releases and include split screen, safe white and black clips, and gamma adjustments. The color correction effect features red, green, and

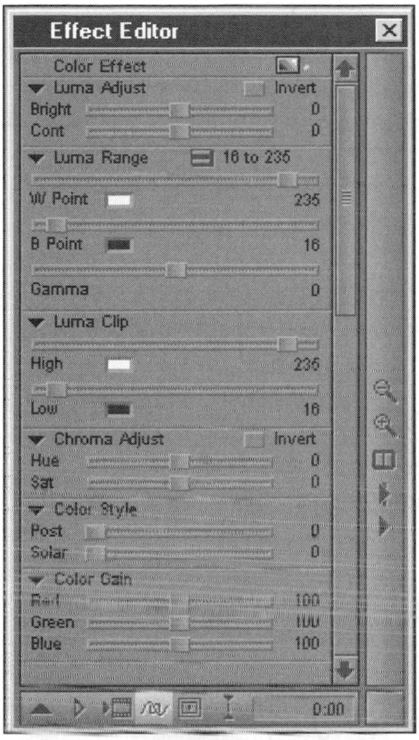

Figure 2.4 Color Effect Interface

blue adjustments—more adjustment than you have with just a straight time base corrector (TBC). With the latest version of Symphony, there has been a breakthrough in functionality that I will discuss in more detail in Chapter 12. If you are primarily an offline editor, you are probably not also a full-time colorist. There are two basic questions about your workflow: Who is running the box and when?

Let's look at a few different scenarios:

Scenario One
You digitize everything for offline and now go back to digitize at high resolution. You set everything to color bars, but there are a few shots that just do not match. You take the time and your best shot at making these simple changes using the color effect, and the client signs off and goes home.

Scenario Two
After digitizing at high resolution, you realize that significant sections require color correction, either because of mistakes in the shooting or just a difference between cameras. Now you have a more serious choice. Is it just the juxtaposition of certain shots? See Scenario One. More likely, entire tapes need to be tweaked.

In this case you could create a color effect template and apply it every time that tape is used. You could redigitize using an external color corrector. Some inexpensive digital color correctors can be bought or rented just for this purpose. You need to digitize each shot and evaluate it or, after some experimentation, you can find a setting for each tape that is a best light, or tape-based setting. This calls for a specialized skill that is only developed over time. You might consider finishing on a Symphony system with color correction.

Scenario Three

You know you are going to be using lots of different cameras under many different locations and lighting set-ups, and you plan for tape-to-tape color correction at a professional facility after the final show is output. If you are planning on this method, you should also plan to master on a digital tape format so there will be no further loss of quality during the tape-to-tape transfer. Again, this is where a Symphony system would allow you to keep this stage of the work in-house.

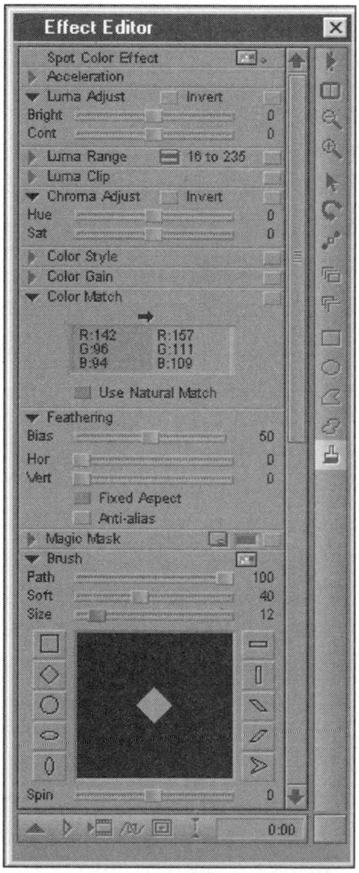

Figure 2.5   Spot Color Effect on Symphony Combines Paint and Color Correction

You may have a colorist on staff, hire a freelance colorist, or have an online editor who has some experience in matching scenes. By avoiding the tape-to-tape stage and keeping it all nonlinear, you can cut costs, finish faster, and keep the final stage all in one room. You can sell this service to your clients for a rate lower than a standard tape-to-tape color correction session or keep all of your rates the same and pocket the difference.

A Symphony tied to a media-sharing solution like Unity will allow the colorist and the editor to work on the exact same frame of video at the exact same time. You now have the flexibility to work in parallel and merge the color correction information with the final version of the sequence at the very last stage. This parallel workflow with shared media solves the most difficult color problems with the talent of an expert and does not affect the editing schedule.

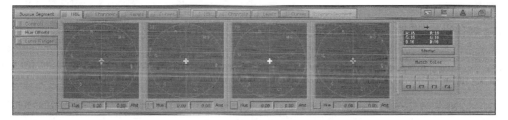

Figure 2.6 Hue Offset Interface on Symphony's Color Correction Mode

RENDERING

What would a digital nonlinear project be without rendering? A lot faster. Part of the strategy of this medium is knowing what to render and when. With the invention of ExpertRender™, much of the guesswork has been eliminated. By choosing an ExpertRender in to out, you have the benefit of the system itself analyzing the sequence for you. Most times ExpertRender will show you the least amount of effects to render and allow you to take better advantage of the large amount of real-time capabilities. The times that you may disagree with the Expert, you can easily override the choices and find perhaps a better way to render the least.

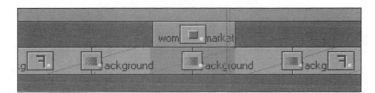

Figure 2.7 ExpertRender Analyzes and Selects Only What is Required to Render

Are there times when you don't need to see all the layers? Take advantage of them and move your video track monitor to view just the minimum. Sure, you need to see the text on V3, but not the color gradation on V2, so render the dissolves on V1 and leave the other tracks unrendered. The dissolves play fine, the titles come up in all the right places, and they are spelled right. When possible, you can Control or Command-click on a video track monitor and turn on video solo to view just one effect, which will now play back in real time. Of course, you need to explain what you are doing and use your best judgment. Determine how much your client can see (or not see) before they start to get cranky and you have to stop and render everything to show them what it will really look like.

Think about the kinds of changes that force you to re-render and whether you can make a series of trims and keep everything intact that is already rendered. These trimming considerations are skills that are more advanced. They require you to ensure all the affected tracks are active when you make changes when you have lots of rendered effects. This is why you want to get as far into the project as possible before you start to add the effects. You have enough to keep in mind when you are still refining the opening montage without worrying about the rendered layers at the end of the show.

Just how much do you really need to render? Most people render too much, way too much, and it is all because of misinformation or ignorance. Think about the minimum necessary to render and the minimum to monitor. The less you see, the less you wait. Will the machine render on the fly and show you the effect when you are concerned only about content? This book will discuss rendering in more detail in Chapter 7.

FINISHING

Throughout this process, there needs to be one thing in the back of your mind: How are you going to finish this thing? Are you going to a linear edit suite, the negative cutter, or does it stay "in the box"? Good form means that no matter what the decision for finishing will ultimately be—and the producer will at some point be forced to tell you—it shouldn't make too much difference. This means you need to name the tapes for an EDL, but make the names obvious so that you or your assistant can find them again easily for redigitizing.

The amount of time you spend on effects should be more concentrated on the show that you know will be redigitized on the Avid for finishing. Otherwise, you are just making a visual guide for re-creation by the online editor. This is a decision about the amount of refinement and a focus of time and effort. It is not unusual to spend an inordinate amount of time on the final five percent of a project if you are finishing. This time would normally have been turned over to an online editor and that would have been their sole job. It is this last stage that most beginners completely underestimate.

There are several ways to save time and money at the final output stage. If you can keep the project in-house, then you can reap extra profits. You can charge for the extra hours involved in finishing instead of figuring in the $400 per hour cost of turning the project over to someone else. Investing in the highest quality video boards and fastest drives can pay off for keeping everything in the Avid. But to truly save money in the finishing process requires a sign-off from your client that there will be no further changes (get it in writing!). If the project must go to a linear online, and as long as the client realizes that you can save hours by manipulating the EDL, then sign-off is the key to the true "auto-assemble." If the show is mostly straight cuts and requires many source tapes, this time savings will be significant.

KEEP THE AUDIO

There is no compression of your audio when you digitize, no matter what your video looks like. You can use this audio in the linear edit suite—you just need to get it there. Copying those audio files to a removable drive and carrying it to a digital audio workstation is becoming more popular as it gets simpler. You may also want to hand-carry or copy over a network your AIFF audio files from your offline Macintosh system to a Symphony. An older method is to record the Avid 8-channel audio to a timecode-controlled DAT, put it to digital videotape, or use audio tracks three and four of a Betacam SP. You can lay the audio to the new master tape in the linear suite and then make your EDL video only. You can keep the mix you created in offline and just keep an eye on video levels and effects in the linear suite. Chapter 8 describes this in more detail.

The last important step when you are going to finish on the Avid editing software is the tape itself. You need to black a tape and, unless you have two decks, you need to black the tapes for protection masters as well. You don't really need to black a tape if you know how to use the assemble edit function available on the later versions of the Avid editing software, but it simplifies things marvelously. Chapter 14 discusses blacking tapes.

PLANNING FOR CHANGES

When everything is finished, the product is shipped, and the check is in the mail, you must concern yourself with one final nightmare scenario. What if they come back? Well, if they are bringing new work this is a good thing, but if they are coming for changes you need to be ready. Can you charge for the changes? Whose fault is the typo? (A good reason to get them to type all the text and paste to the Title Tool!) But the real trick is being able to make the client's changes quickly and painlessly, and they will certainly keep that in mind when they are projecting costs for their next project.

How are you going to archive this project? Don't be tricked into thinking that an EDL will do the job. You must back up the project file and all the bins. Even if the software changes, you should usually be able to bring any job forward even if it is a year or so later. Large projects require large backups, and it may be time for you to invest in removable storage like one of the new, fast, removable formats like Iomega, or even a recordable CD-ROM just for the bins and graphics.

Consider also whether you need to back up the digitized media. If you are finishing on the Avid editing software, you should build into the cost of the project the amount of time to back up overnight and the cost of an archive DLT. The DLT technology gets faster all the time and is the only way to realistically back up such huge amounts of data. If you have spent the time to color correct and render and everything is just right, then you should spend a little extra time and save it. If you have a day's warning, you can have the entire project ready to pick right up again by restoring it the night before. The potential downside is that, when the technology changes, it will change video quality faster than anything else. Eventually, video-digitizing boards will change and your new video will most probably be incompatible with the old video if you have upgraded. With luck, there will be a conversion utility, but the quality will never be as good as if you redigitize. Not everything ages as well as wine and cheese.

So plan for an adequate amount of time to review and make changes because there will be more of each in a nonlinear project than in a linear project. Have a plan to back up and restore media and twiddle your thumbs occasionally. Everything takes longer than you think it will, but expect the slowdowns and crash edit times to happen in new places.

Workflow encompasses much more than just input, edit, and output. It is full of variations and possibilities and is guaranteed to be unpredictable. Plan for it and be ready for the places where the changes could cost you money and time. If you understand the process, you will be better prepared to avoid the traps and exploit the advantages of the new technology.

3

Intermediate Techniques

It may come as a surprise that most beginners at Avid editing make the same basic mistakes. I don't mean mistakes caused by the software being too difficult, but mistakes from working hard to grasp some fundamental ideas. They are sometimes crucial mistakes, like not knowing exactly what to back up and then trying to restore a project with no bins. Sometimes it is a subtler mistake, like not using the power of a new tool because "That's not the way I work." You may be missing a huge opportunity to improve your speed and understanding.

I have heard people say that the Avid is difficult to learn and that the interface has a steep learning curve. This is only partially correct. You can be mousing around the screen in only a few hours and really editing by the end of your first day. But as with any professional tool, you want it to go faster and do more and the Avid interface rewards this. The more times you resolve different post-production needs, the more you can use the functions you already know in new and different ways. Very few people use everything the software has to offer. If you have a particular task to perform, there are the tools designed to facilitate that task in a straightforward way. If you need something a little different, there is a lot of room for variations. It is the variations that take the time to learn and are the most rewarding.

The most basic mistakes are made right at the beginning, when editors are still trying to learn how to navigate through their material. There is a lot of translation going on between where they want to go, how they used to do it, and the two or three techniques they know how to use. They end up settling for the dog paddle before they have mastered the breaststroke, the crawl, the backstroke, and the sidestroke. They will always poke along unless they unlearn the method that wastes energy and, let's face it, gets you there without much style.

USING THE KEYBOARD

What you may have guessed by now is that I am referring to the overuse and abuse of the mouse or trackball. This is where you should start to improve your technique. When you first learn, use the most obvious way, the mouse. This helps you get over the beginner's problem of trying to remember what you want to do next and where it is on the screen. After this beginner's stage, you instantly forget how magical it all is and want to go as fast as possible. Then you must cast down the mouse! Use your keyboard!

The best favor you can do for yourself is to force yourself to use the edit keys and keyboard equivalents as soon as possible. If you haven't put the colored keycaps or stickers on your keyboard yet, you are missing a whole world of speed. Look at the Ctl/Command key equivalent for the functions you use the most and think up funny little ways to make them stick in your head. Ctl/Command-Z to undo and Ctl/Command-S to save should be comfortable before your first day is over. Then start to use Ctl/Command-W to close windows and Ctl/Command-A to select all. Use Ctl/Command-B to go to Capture Mode and, as an ongoing project, memorize the Tools menu. Mark in and out should be mostly done from the keyboard so you can keep your material rolling and mark on the fly. Trimming can be done several ways from the keyboard.

You don't want to give yourself a repetitive stress injury, though, so always try to make the environment as friendly as possible for your wrists. Get the keyboard at the right height, get a wrist pad if you need it, and give your wrists the rest time and exercise they need to keep functioning. Use the extra time gained with these keyboard techniques to watch the sequence back one more time and think.

Custom Keyboard

In Media Composer and Symphony, there are many ways to personalize the keyboard. With Xpress, you can take a keyboard that has been customized on a MC/Symphony system and, if the features exist on Xpress, it should work. I don't recommend any special keyboards since I believe they should all be created organically from observing the functions and keys you use the most. You can take any button from the Command Palette and put it on any key (button to button). I recommend learning the colored keycaps and only modifying the function keys or the shifted functions of keys that make alphabetic sense to you. Open the Keyboard Setting and hold down the Shift key. You have a whole new keyboard to create! You can put the Render button on Shift-R or the Subclip button on Shift-S.

Some functions on the pull-down menus do not have keyboard equivalents or buttons. Mapping a pull-down menu to a key is now a single button in the new Command Palette for versions 2.0/7.0 on the Mac and 2.0/8.0/1.0 on NT and later.

In the recent versions, here's how to map a pull-down menu to a key:

- Open the Keyboard Setting and the Command Palette.
- Click on the Menu to Button reassignment button on the lower right.
- On the Keyboard Setting, click on the key you want to map. Hold down the Shift key if you want it to be a shifted function.

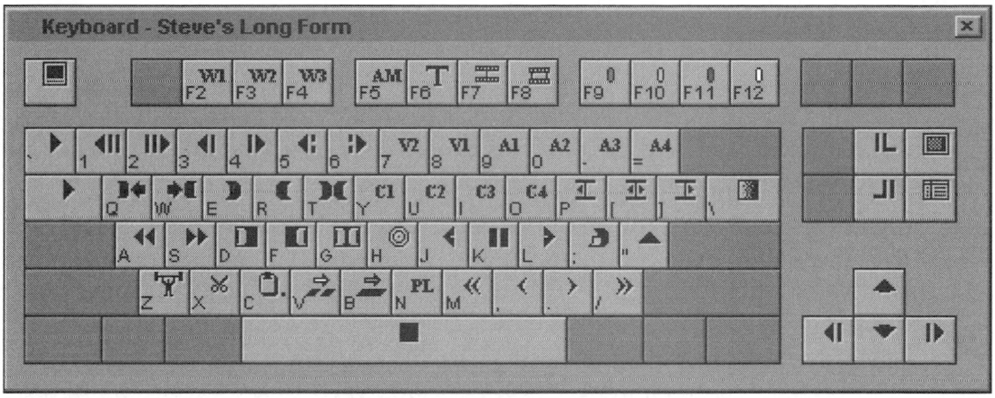

Figure 3.1   A Customized Keyboard

- Choose the menu you would like to map from the pull-down menu choices.
- Save your settings when you are done.

Pull-down menus must be mapped through a special combination of keystrokes in older versions of the software. Here's how you do it:

- While looking at the open Keyboard Setting, hold the Option and Command keys down at the same time; you will get a cursor that looks like a small pull-down menu.
- Click on the function key in the Keyboard Setting where you want the function to go.
- Choose the pull-down menu you want to map, and the initials representing that function will land on the function key.

You may have made the same mistake I did and didn't learn to touch-type in high school; who knew about e-mail then? Once you feel comfortable with the screen real estate, push yourself to resist the mouse and keep your hands on the keyboard!

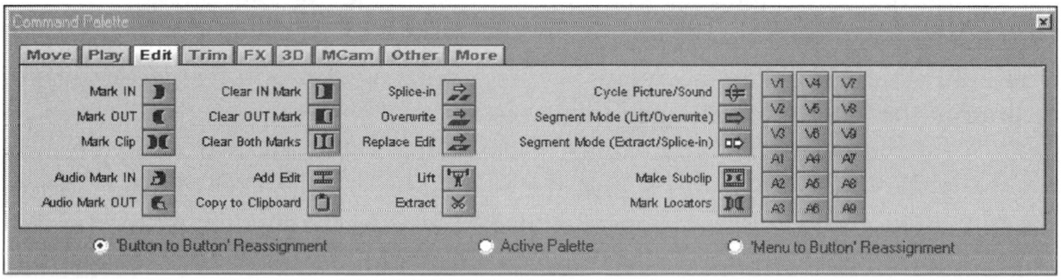

Figure 3.2   The Command Palette

If you have a custom keyboard, be aware that it works best for the version of the software you were using when you created it. If you go to an earlier version of the MC/Symphony, your keyboard, like all User Settings, may have features included that do not exist in the earlier version. You can usually go forward with Keyboard Settings, but this has been known to occasionally create odd, unrepeatable problems. It is best to take a screen shot of your Keyboard Setting (use a shareware screen capture program to outline the part of the screen you want as a PICT file), print it out, and make the custom keyboard again.

J-K-L

Use keyboard control for shuttling. Some love the MUI or the AvidDroid because it feels like what they are used to. Those devices are better than using the mouse to shuttle or drag through the Timeline to view material quickly. Programming the extra function buttons can give an edge to the film-based editor with a yearning for the flatbed systems. Some like it especially for shuttling tapes during the initial viewing stages of the project. But if you don't have access to that extra hardware, the keyboard can be just as fast after the material has been digitized.

Getting through your vast amount of material quickly and then fine-tuning the details can all be done with the J-K-L keys. After you get adept at using the Jog buttons to fine-tune, stop using them! Hit the L key and move forward at 30 frames per second (fps), then hit it again and speed up to 60 fps, and then again on up to 90, 150, and 240 fps. (PAL numbers will be multiples of 25 and film is 24.) Eventually the sound will cut out (thankfully) above three times normal speed. Hit K or the space bar to pause. Now use the J key to go backward with all these different speeds.

Soon you will find yourself cooking through the material at double or triple speed and following the script. Surprisingly, you can understand what they're saying and can work consistently at the higher speed to fly faster through the material without the distracting rollbars on the screen that you would see with tape.

This is a linear way of moving through the material. J-K-L is an example of how the computer has improved on an old technique by just making it easier and faster. Ideally, you type in your timecode logs and go to the correct place instantly. Take advantage of the random access powers. You search back and forth once you get there; however, if you really need to listen and watch the material, there is no replacement for just shuttling through, especially if you were not the person who logged the shots.

The true beauty of J-K-L comes when you want to make that final editing decision. Holding down the K and the L at the same time will "scrub," move slowly, with audio at 8 frames per second (5 fps PAL). You also get the familiar analog sound effect where the pitch drops and it sounds like Satan has possessed your talking head. Most people like this sound, especially if they are used to editing audio in an analog environment.

J-K-L for Trimming

Using the keyboard, you can move incredibly fast and slow without taking your fingers off the keyboard. Just wait until you get to the left hand. First, let's take J-K-L to its most powerful (and potentially dangerous) use—trimming. The first time you try this there is a little surprise—while watching the show play back, you are trimming! There are two important benefits to this. Don't scrub and then mark the points for your edit. Make the edit first and then scrub the trim. You will be able to make the fine decisions with the shot already in place. Second, if you mistakenly make a clip way too short, say by a paragraph, you can lengthen it and listen to the audio at 90 fps at the same time. This is an excellent way to hop through the sequence, one transition at a time, and tighten everything just a bit before moving on to the next level of refinement—always keeping the hands on the keyboard.

Scrubbing and Caps Lock

You can achieve a choppy, digital scrub sound by holding down the Shift key and using the Jog or Trim keys. If you like this better than J-K-L, it is precise and still has its place. This technique should be a last resort at the very last stage of determining a trim, not the first choice. It is good to know first because you can always fall back on it, but you will soon see as you speed up with the rest of the interface that, as a main technique, it becomes a dog paddle.

There are some other restrictions with using the Caps Lock key. First and most important, the Caps Lock key takes up a lot of RAM; a lot more than you would think. Maxing out the RAM may cause unusual errors or keep you from seeing certain real-time effects when they are definitely there, like many dissolves in a row at AVR 75. Now that systems are shipping with triple the RAM of the original configurations, this isn't as much of a problem. It is something to consider if you have an older system and less RAM. Also, RAM is much cheaper than it used to be. If you can, put this book down, go to the phone, and order another 64 megabytes or so. This is the cheapest upgrade possible.

The second problem is that leaving the Caps Lock key down is incredibly annoying. OK, you've escaped from having to hear high-pitched voices screaming backward when you abandon tape, but now you have blip, blip, blip all day every time you change position in the Timeline. Stop that! Turn it off!

There, did I sound enough like your mother to get your attention? Hold down the Shift key when you need it; when you don't, it automatically turns itself off when you remove your finger. And start using J-K-L.

AUDIO MONITORING

Many beginners don't see the connection between what they are viewing and why they can't hear the sound anymore. Watch the small speaker icon next to the audio tracks in the Timeline. The solid, completely black icon means that you are monitoring that track.

Being able to turn audio track monitors on and off selectively means you can concentrate on just the sound effects or make five versions in separate languages and monitor one language at a time. But it also means you always need to keep an eye on which tracks are being monitored because they could be different from the tracks you are editing.

Look closely at the monitor speaker icon. On older systems one of them will be hollow, represented just by an outline instead of solid black. On the newer systems, some icons will be a bright green (which is much easier to see with the high-resolution monitors). This is the track that you monitor at the off speeds. You can move this icon to another track by Alt/Option-clicking on a solid monitor icon.

With older systems, you are restricted to listening to only one track at a time when moving slower or faster than 100 percent normal speed, or moving backward. There was a hack for a few versions before MC 6.5 that allowed you to monitor more tracks, but this was eliminated because it caused people to occasionally have underruns on slower systems that couldn't actually always play everything at once. The newest versions on NT (3.0/9.0/2.0) will allow you to monitor a maximum of two or eight channels while using these off speeds. The place to change this monitor choice is under the Audio Settings.

If you are only monitoring one channel for off-speed scrubbing, then you can be moving forward at 30 fps listening to music and voice and then when you

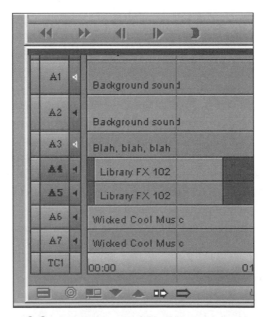

Figure 3.3   Monitoring and Scrubbing Eight Channels

reverse, you are left with just voice. In general, this is what you want when you are doing fine audio cutting. Having all eight audio channels playing at once is distracting when you just want to find the "blip." Even though you may have eight channels potential for monitoring, you can turn them on or off as desired. The eight channels is a maximum and is really only going to be useful when you have lots of separate sound "stems" or are checkerboarding dialog across several audio tracks.

Scrubbing Multiple Audio Tracks

Using some of the increased bandwidth of the NT systems, Avid has been able to increase the amount that audio channels can be monitored at low and high speeds as well as backward. In the Audio Settings of NT releases 3.0/9.0/2.0, there is now a choice between Minimum and More Audio Tracks. With the Minimum Setting chosen, you will be able to monitor two channels all the time at all the various monitored speeds (audio still cuts out above 3X normal speed). The way you can tell this mode is on is that you will have two of the bright green audio monitors in the Timeline to distinguish those tracks from the normal black audio monitor icons. Any track can be made to be an off-speed monitor by Alt/Option-clicking on the monitor icon. One of the biggest workflow improvements of having all these tracks monitored is when you are using J-K-L keys to fly through lots of material and you have checkerboarded the dialog. With one sound byte on V1 and the next on V2, you can monitor something that sounds like the finished audio at all the off speeds when you are trimming and continuing to tweak the sequence.

THE OTHER HAND

Now let's cover the other hand. The E and the R keys should be the main resting place to mark in and out on the fly (or I and O if you are reversing all this). Using the keys to mark in and out allows the video to keep rolling while you continue to search and change your mind. Combine this with Make Subclip mapped to a function key and you can fly through material and organize at the same time. Ctl/Command-4 on MC/Symphony takes you from the bin back to the source window to subclip, and you can subclip and name an entire bin without taking your hands away from the keys.

Alternatively, use the thumb for the V or B keys to actually splice or overwrite the material into a sequence. If you are working from long source clips with many different shots in one master clip, like a film transfer, you can literally type your show together. Use the mouse mainly for loading clips into the source monitor (although the Return key will do that, too) and for accessing the graphic functions of the Timeline.

MULTIPLE METHODS TO SOLVE ONE PROBLEM

There are ways to do the fine adjustments and still get the analog scrub sound, but with the mouse. It's called the mouse jog and it's a finer scrub than J-K-L. Hit the N key to enter the mode and the space bar to exit. This brings me to another fundamental tenet of using the Avid systems: You can use multiple methods, one after the other, to fix a problem. Editors who consider themselves novices may have acquired one or two methods that they use under all circumstances. The more experienced editor uses a method of trim to get close as fast as possible, depending on the particular Timeline view and level of sequence complexity. The editor then reanalyzes the problem and easily switches to another method to put the final polish on the fine points. The "right" way is the way that accomplishes what you want in that particular situation in the fastest way possible.

NAVIGATING NONLINEARLY

Another beginner's challenge is not to think linearly when jumping large amounts of time. Editors make a compromise and move at the speed of human comprehension in order to grasp a particular point in the material or listen to the performance. But what if you just want to get there as fast as possible? I see beginners actually dragging the Timeline through their material or sequence to get to the end! It's random access; get random. Jump to the end with the End key and jump to the beginning with the Home key. (These can be mapped in version 7.0 and later.)

FAST FORWARD AND REWIND

The Fast Forward and Rewind buttons are useful if you just want to jump to the next or the previous edit, but these buttons are usually left to the User Setting default of being "track sensitive." This means that the Fast Forward and Rewind buttons jump to the next edit that uses all the tracks that are highlighted in the Timeline. If video track 1 is highlighted, then you jump, in a sequential way, from cut to cut on video track 1 only. Turn on all the tracks (Ctl/Command-A with the Timeline highlighted). Now you jump to every edit where all the tracks have a cut in the same place. No straight cuts on all your tracks in the same place? With all your tracks highlighted, Fast Forward jumps all the way to the end of the sequence. That can be confusing since there is no easy way to get back to where you were in the Timeline. No undo for jumping to the wrong place! You can reverse this default Composer Setting instantly by holding down the Alt/Option key in combination with the Fast Forward or Rewind keys. If you find yourself

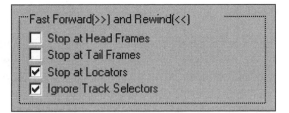

Figure 3.4   Fast Forward and Rewind User Settings

jumping to the end of the sequence a lot by accident, then in MC/Symphony, you can change the Composer User Setting so that it jumps to every edit regardless of which tracks are highlighted. You can change the settings to jump to every locator, too.

USING THE TIMELINE

Modifying Fast Forward and Rewind still misses the larger point, which is that you don't need to step through lots of edits just to get through the Timeline. If you want to get near the end, just click there! But more important, you need to be able to see where you are going. Learn to change the scale or view of your Timeline quickly and easily.

Version 7.0 of Media Composer and all versions of the Avid Xpress and Symphony incorporate a drag bar to resize the Timeline. This is marvelous and intuitive. Drag the slider to the left and the Timeline compresses; drag it to the right and it expands. You have fantastic fine control and can change the view by large amounts quickly with the same function. But don't neglect the Focus button (H), keyboard equivalents (Ctl/Command-[, Ctl/Command-], and Ctl/Command-/), and Zoom In (Ctl/Command-M). Some of these functions are on the MC/Symphony only because Avid Xpress is purposefully made simpler to use by reducing the amount of keyboard equivalents. The Focus button (H) is especially useful for troubleshooting small problems like flashframes. Ctl/Command-M allows you to drag a box around the area in the Timeline where you want to zoom in. This technique is very useful for pinpointing a segment that needs more refinement.

There are so many easy ways to change the scale of the view in the Timeline because it is vital to using the power of the system. The Timeline is not just a pretty picture. It is an important tool for navigation and should always be sized to fit your needs for that moment. Zoom in to do some fine trimming, zoom back a little and look at the whole section, then zoom in somewhere else to fix a problem. You should be considering the scale of the sequence that your Timeline is displaying at every stage of work.

JUMPING PRECISELY

Even though you may see precisely where you are going, you may not always get there the fastest by just clicking the mouse. You need to combine the power of the random access navigation of the Timeline with the precision of the Fast Forward and Rewind keys. How can I jump huge distances in a single bound and still end up on the first frame of the cut? There are a series of modifier keys without which life as you know it could not exist. The most important is the Ctl/Command key. When you click in the Timeline and hold down the Ctl/Command key, the blue position bar always snaps to the first frame of any edit on any track. It also snaps to marked in or out points.

If you hold the Ctl/Command key down while dragging your cursor through the Timeline, it snaps to the head of every video and every audio edit. If you click anywhere in the Timeline with the Ctl/Command key held down, you are guaranteed to land at the head of the frame or a marked in or out point and can eliminate missing a frame here or there and creating flashframes. You must still combine this with the Timeline zooming techniques, especially with a very complicated sequence. You may be surprised that you have snapped to the audio edit on track 7, which is three frames off from the video edit on track 1 you really wanted.

Important Modifiers

Here is a list of the most important modifier keys and what they do in the Timeline:

Ctl/Command
Guarantees snapping to the head of a frame when dragging. This is the standard procedure in Xpress and does not need a modifier key.

Alt-Ctl or Option Command
Allows you to snap to the tail or last frame of an edit as you drag. This happens by just holding down the Alt or Option key alone in Xpress.

Alt or Option
Gives you fine dragging control even though you may be zoomed way out in the Timeline view. There is no equivalent in Xpress.

Shift-Ctl or Ctrl
Note this is the only use of the Ctrl modifier key for dragging segments on the Macintosh versions. This key combination will drag a segment vertically without snapping to any marks or existing segments around it. This technique is perfect for multi-layered effects. Although just the Ctrl or Command keys will allow

you to do something close to this, it is not as exact and may cause you to snap to an edit hidden on audio track 9. There is no equivalent on Xpress.

I want to point out the most important key to use when using the Timeline in a true nonlinear way.

Lasso with Modifiers

Holding down the Alt key on NT or Control key on Mac allows you to lasso any transition on any track and go into the Trim Mode at a specific place. In combination with the Shift key, you can make multiple selections easily and, in a graphic way, extend the use of the Timeline to get exactly what you want.

THE IMPORTANCE OF THE LAST FRAME

Avid editing systems work from the film model that insists that every frame exists and is important to duration calculations. The video model actually allows you to have an outpoint and an inpoint on the same frame of timecode. How can two shots use the same frame on the master tape? Don't ask, in linear video editing they just do, and video people are occasionally confused when the Avid editing system counts every frame separately. Mark an in and out on the same frame. What is the duration? One frame. If you are looking at a frame and mark it as an inpoint, then say "Go 15 frames from here" by typing "+ :15"; when you mark out, you will have a duration of 16 frames. You have said, "Take this first frame and 15 more," which makes perfect sense to a film editor and seems suspiciously like a bug to a video editor.

There are two ways to deal with this fundamental difference. If you fast forward to the beginning of one edit (or Ctl/Command-click in the Timeline) to mark in and then fast forward to the next video edit to mark out, how many frames do you have? You have one too many. You have included the first frame of the next edit. If you do an Alt-Ctl or Option-Command-click, then you are at the last frame of the shot you want (or just back one frame). Your inpoints and outpoints are exactly equivalent to the duration of the shot.

There is an easier way to do this, of course, and that is using the Mark Clip button, which is also track sensitive. This technique avoids the entire problem by just taking the first frame and the last frame of the edit you are working with. Leave the next edit alone!

TRIMMING

Beginners bring linear thinking to the trimming process as well. This is the biggest potential mistake you can make, short of deleting all your media. The first place I see this is when beginners misuse the Match Frame button. I would like to rename this the "Fetch button" because the Match Frame name is too close to the function that linear tape editors have been using since the beginning of computer-controlled timecode editing.

The traditional tape method is to get the edit controller to find the same frame on the source material as where you are parked on the master tape. The source tape cues up, you adjust the video levels to match what is already on the master tape, and then you lay in a little more of the shot, usually a dissolve or another effect. You can do this in the Avid as well. This is logical and simple, but misses the point. Every master clip that you add into the sequence is linked to the rest of the digitized material. You don't need to go get it because it is already there. Think of the extra digitized material as always being attached to every edit in the Timeline all the time.

The best reason to use the Match Frame button is to call up a shot to look at the whole thing or to make a motion effect. It should be used for reviewing material, not as an integral part of the trim process. Used the wrong way, Match Frame is another dog paddle.

Trim vs. Extend

Going into the Trim Mode is the more sophisticated, but potentially more confusing, way to extend a shot. It allows you to do more in a complex situation. I personally think that the Trim Mode is the heart of the Avid editing systems' functionality. Master trim and you have mastered the system and changed the way you think about editing forever. Trim is creative, not just corrective.

The Extend button in MC/Symphony is the simpler, more straightforward way of extending a shot. Extend can also be faster than trimming. Mark where you want the edit to extend to, turn on the tracks you want to extend, and then hit the Extend button. It will not knock you out of sync because it is a center trim. It trims both sides of the transition simultaneously. I find it most useful for mechanical trims that go to a direct and easy-to-mark point. Any finessing and I go to the Trim Mode.

Mastering Trim

The main reason that trimming is so much better than just extracting the shot and splicing it back in is that you have the immediate feedback of seeing the shot in context. Fixing a shot while it is in place gives you the instant feedback that is so important when using a nonlinear editing system. The one issue to keep in mind when expanding your use of the Trim Mode is staying in sync as much as possible. Now obviously, there are times when you want to go out of sync, for cheating action or artistic purposes, and I'm not talking about that. I'm referring

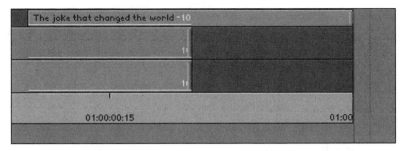

Figure 3.5 Timeline Sync Breaks

to the skill of understanding the relationship between what tracks are high-lighted and what kind of a trim you are doing. Some people get so flustered the first few times they try trimming with sync sound that they abandon it alto-gether and invent elaborate workarounds that are easier for them to understand Lots of energy, not much style. This is one of those skills that film editors (those who have actually touched celluloid) have over video editors. It is pretty hard to knock yourself out of sync with a tape-based project, so thinking in terms of maintaining sync is quite foreign. But film editors must learn that whenever they add something to the picture—a trim or a reaction shot—they must add an exactly corresponding amount of frames to the sound track.

With the extra power of the nonlinear world (and film was the first nonlin-ear editing system!), there is the responsibility of keeping track of sync. The Avid editing systems do a pretty good job of telling you if the video and audio you digitized together or auto-synched together (matching sound and vision from separate sources after digitizing) have lost their exact relationship. They are the white numbers called sync breaks. I think of them as a silent white alarm that, when I see them ripple across my Timeline, tell me I have most probably made a mistake. The only time I want to see sync breaks is when I have cheated action or I am dropping in room tone.

Staying in Sync

The best way I have discovered to think about trimming is to imagine moving earlier in time or later in time to see a different part of the shot. Coincidentally, as you move earlier you may be making a shot longer or shorter. Any trim that adds or subtracts frames—any trim that is on one side or the other of the tran-sition—changes the length of that track and must have a corresponding change on all of the other tracks in the sequence. Not all trims change the actual length of a sequence, but the ones that do—the trims on one side of the transition or the other—knock you out of sync if you don't pay attention. This means you must look to the tracks that are highlighted when you decide to add a little video.

Don't make the beginner's mistake of thinking that just because you are adding a few more frames to lengthen an action, it is a video-only trim. All the sound tracks must be trimmed if you make the sequence longer or shorter in any way.

Sync Locks

The easiest way out of this dilemma is to turn on the sync locks. But this is at best a temporary solution. The sync locks allow the system to resolve certain situations where you tell it to do two different things: Make video longer and don't affect the sound tracks. The system adds the equivalent of blank mag— silence—to the sound track. This may be safer than trimming and accidentally adding the director shouting "Cut!," but it will also leave a hole that must be filled in later. Blank spaces in the sound track are really not allowed! You will find yourself having to return and add room tone or presence so the sound does not drop out completely.

There will be a time when you tell the system conflicting things. You tell it to make the video shorter, don't change the audio tracks, and stay in sync. This is beyond the laws of physics. In this case, the system cannot make the decision for you where to cut sound in order to stay in sync, so it will give you an error beep and do nothing.

Sync locks work best if the majority of your work is straight assembly with little complex trimming. It is very effective, however, when you are locking a sound effect track to a video track. Sync-locking multiple video tracks together to keep them from being trimmed separately will help you keep from unrendering an effect.

There is a potential situation where you will be cutting video to a premade sound track. The video and audio parts of the sequence do not give you sync breaks when you change their relationship. Here, you must be even more careful to be conscious of maintaining sync. Don't fall into the trap of thinking that you can knock yourself out of sync now and later, when you get a chance, go back and fix it. Believe me, by the time you get the chance to go back, you will have created a situation that takes much longer to fix than if you did it right in the first place.

Trimming in Two Directions

Trimming in two directions, an asymmetrical trim, is a subtle and very powerful technique. You may have a situation where you must trim a video track longer, but then add extra material to a sound effect or music that would keep you in sync, but ruin the edit. Here, keep in mind that just because you are adding video on V1, you can still trim the black before the clip on A2. Both tracks get longer, but one adds more digitized video, fixing a visual problem, and the other adds more silence, keeping the two tracks in sync.

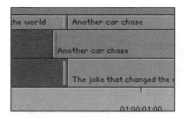

Figure 3.6 Trimming in Two Directions in the Timeline

Slipping and Sliding

Beyond the basic situations, most of the difficult sync problems can be fixed by two methods. The first method is to ignore the fact that trimming a shot must make the sequence longer and trim in the center of the transition, basically not affecting the sync, and then use the Slip function. Slip and slide are not features in the Avid Xpress. Many people have a difficult time grasping the Slip function because it is so tied to the nonlinear, random access concept. In MC/Symphony, you can enter the Slip Mode by multiple methods. You can double-click with a segment arrow if there is no effect or double-click while in the Trim Mode as long as the Timeline view allows you to see a black arrow cursor. You can also get there by lassoing the entire clip from right to left. I generally get there while still in the Trim Mode by double-clicking because I use it in the following method as a second step in a difficult trim situation.

Think of slipping as a shot on a treadmill. The shot slips forward or backward, showing an earlier or later part, but the place in the Timeline never changes. A slip will change the content of a shot by revealing new material but leave the duration of the shot and location in the Timeline the same. Because you usually have more video linked to any shot used in the sequence, you can slip that entire shot back and forth.

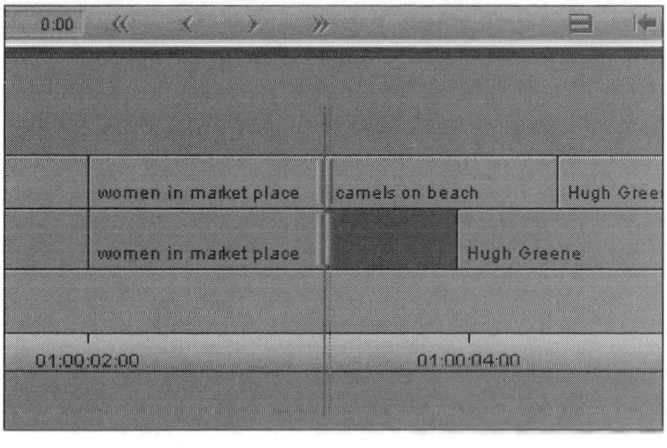

Figure 3.7 Selecting the Transition to Lengthen

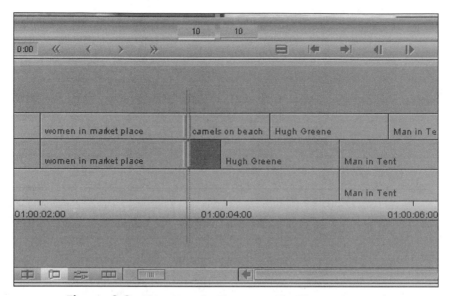

Figure 3.8   Trimming the Transition Ten Frames Forward

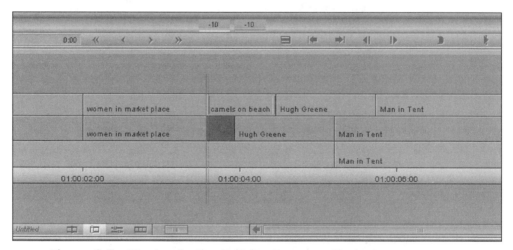

Figure 3.9   Slipping the Shot Back Ten Frames to Get the Same First Frame

So if you trim the beginning of the shot ten frames as part of a center trim, you can slip the shot back into position so that it still starts with the same frame. If the first ten frames of shot B are important, then slip them back into place. Your center trim moves the frames viewed in the A and B shots of a transition to be ten frames later. Although shot A gets longer and shot B gets shorter, the length of the sequence is not affected and the sync is not disturbed. Shot B has gotten shorter, but after you slip, it still has the same starting frames. I have worked with producers who have edited their programs on the MC/Symphony for years who had never seen the Slip Mode! Although it seems complex at first, it is truly a powerful tool used in the right place.

The corollary to the Slip Mode is the Slide Mode. It moves the shot neatly through the sequence by trimming the shots on both sides. It affects the location of the shot in the Timeline, but not the content or the duration. It is a good alternative to dragging a shot with the segment mode arrows if you are making a smaller change, but not nearly as useful as the Slip Mode.

Alt/Option-Add Edit

The other main technique to deal with difficult trim problems and maintain sync, especially with many tracks, is the Alt/Option-add edit. It is very easy to just lasso all the transitions around the area you want to trim or, with the Timeline window active, hit Ctl/Command-A (select all), turn on all the tracks, and then go into the Trim Mode. This can be unpredictable if you have zoomed in on a particular transition to work on it closely and the next edit on V2 is over ten minutes away. You don't really know how you are going to affect this other track, V2, unless you zoom back out to see. You might accidentally add more video to a shot on V2 that you think is already perfect or unrender an effect. The best method for trimming multiple tracks, but with empty spaces on some tracks, is to go into the Trim Mode and use Alt-add edit or Option-add edit. This automatically adds an add edit in the blank tracks and gives you a place to trim where it would normally be just black filler.

The Alt/Option add edit gives you the handle to stay in sync when trimming multiple tracks. You are left with add edits in the filler—fake transition points—but you can get rid of them by choosing "Remove Match Frame Edits" from the Clip pull-down menu. When you have selected the add edits in the Trim Mode, you can also hit the Delete key. But my advice is just to leave them.

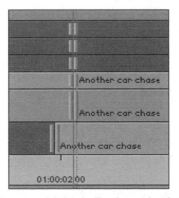

Figure 3.10   Trimming Multiple Tracks with Alt/Option-Add Edit

They provide a continuing ability to make changes at that point in the sequence and are a good indicator that, at some point in the future, you can pinpoint exactly where you knocked yourself out of sync. They won't show up in your edit decision list (EDL) unless you put a dissolve on them—unlikely for add edits in black.

So the most important aspect of trimming is to be aware of when you are going to change the length of the sequence by trimming on one side or the other of a transition and which tracks will be affected. When you grasp these points and overcome the fear of going out of sync, you will have a much more powerful tool and feel much more comfortable with the workflow concept of refine, refine, refine.

ORGANIZING

One thing that working with a computer forces you to do is organize. If you don't have a plan from the beginning that is easy for you, you just won't do it and will find yourself relying on the frame view to find shots. Although this may make it easier for clients to put their greasy fingers on your monitor and shout "That one!," it is slow. The bins that come from the telecine need to be named for scene and take, but after that, everything follows the standard script notation. In a traditional film edit room, organizing has more to do with tracking the physical film and keeping the right scenes ready for cutting. In a nonlinear edit, a film project must keep everything from a scene together. Documentaries or other formats where the form takes shape during the edit must live or die by good organizing of shots in a computer-based edit. I once edited a show with 200 sound bites and no script. In order to use the tools that are available for finding shots, you need to enter the information in a way that allows you to search for it easily.

There are two things to keep in mind when creating bins: tape name and bin size. Generally, the tape name is most important when you are starting to organize since this is what you are initially handed from the field. Tape names should have some criteria for allowing you to trace back shots. For instance, with day and location coded into the number used for the tape name, you can more easily follow a series of shots to a common starting point. Also, as mentioned before, you can start with a tape named bin for digitizing and then organize based on content.

If your bins are too large, however, you are defeating a lot of the benefit of the organizing. It takes longer to open and close large bins, and once they are open, it takes longer to find exactly what you need. Since you will most likely use the Find Bin button frequently, you want it to be opening small bins to speed up the retrieval process. To use the Find Bin function, you must have that bin listed in the project window (or in a folder in the project window); it needs to have been opened once in the project. Also, you can go directly from the sequence window to the source bin by using Alt-Find Bin or Option-Find Bin. The new type of folder-based hierarchy in 2.0/7.0/1.0 and later will make the Find Bin feature more successful. Again, Match Frame is also a useful tool for calling up a shot if you are not interested in the bin. It calls up the source clip regardless of whether the bin even still exists! All Match Frame needs is media on the drives.

THE MEDIA TOOL FOR EDITING

The Media Tool is a good way to find shots, especially if the bin has been deleted, by sorting or sifting and then dragging the shot to a new bin. If the media is online, you can get it through the Media Tool. Most people use it for media management, but it is an easy way to find hidden shots quickly by searching across projects.

On a very large job, or a system with lots of media, opening the Media Tool can be a time consuming process causing some people to avoid it completely. That is why later versions (everything after 2.x/7.x on Mac) have a couple of new choices before opening. Now you can choose exactly what drives or what projects you will search through. The smaller amount of media that is searched, the faster the Media Tool will open. Opening the Media Tool causes the system to read the Media Databases from every active MediaFiles folder and load that information into RAM. That can take drive access time and potentially use up much of the available RAM (if you have hundreds of thousands of objects). Of course, if you are really desperate to find a shot, you can just hit "All Drives" and "All Projects" and load the entire Media Database. Once loaded completely, all searches in the Media Tool will be instantaneous for the rest of the session or until you shut down the computer.

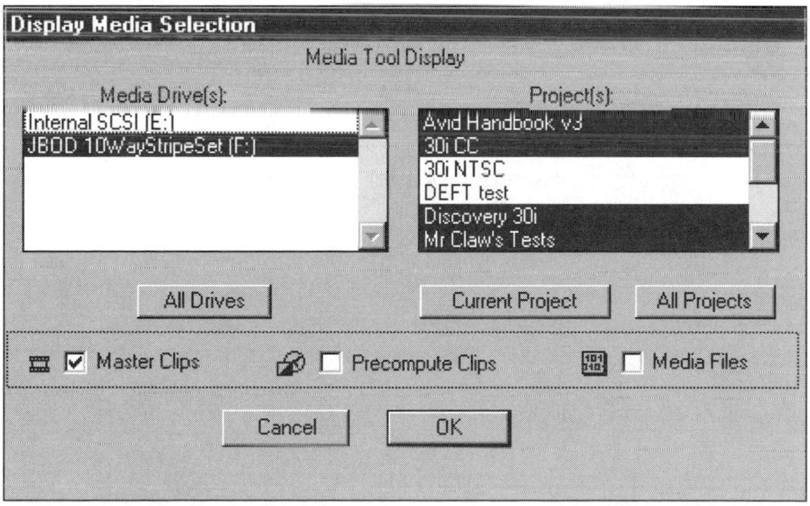

Figure 3.11 Media Tool Choices for Finding Shots

Early in the process of organizing a new project, if the Media Tool is not too large yet, I leave it open to avoid having to keep opening up a bunch of separate bins when looking for a shot. In the Version 7.x Macintosh operating system and earlier, when there are too many objects in any window, the scroll bars on the sides of the window disappear. If you are running this older version of the Macintosh operating system, then the scroll bars will disappear on the Media Tool when you have too much material online. Think of the Media Tool as the largest bin of media you can have at one time and, when it gets really large, only open it when you must.

To search more effectively, make use of the customized columns in a bin and try to standardize the way you enter particular information. If you create a custom column called Shot Type and enter both "Wshot" and "WS," you will have a harder time searching. Once you have entered a particular notation in a custom column, whenever you go back to that column and Alt-click or Option-click, you will get a list of everything you have entered. You can choose a common entry from this temporary menu. This lessens the chance of spelling errors and speeds up data entry.

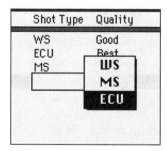

Figure 3.12   Reusing Standardized Data with Alt/Option-Click

So far we have discussed stylistic techniques, and the benefit gained is primarily in speed, efficiency, and making you look good. But a few common techniques, if not followed regularly, can create real technical problems. Chapter 9 describes some of these problems, but let's cover a few that can be easily prevented by just following directions.

BACKING UP

If you are a beginner to the computer, you may not realize the seriousness of backing up, but first imagine the cost of losing a day of work. Then imagine losing the entire project. Until you lose your first project, you might not back up on a regular basis. The fact that for about US$100 and five extra minutes you could have easily saved all the project information is a very compelling argument. I back up every day, but not to the same disk! Use a separate removable disk for every day of the week, and you have seven chances to find a usable version.

What should you back up? It is usually difficult to back up all of your digitized media in the middle of a project. But with a DLT and backup software that allows you to back up invisibly while you edit (like Avid/Mezzo Archiver), backing up media isn't out of the question. Backing up digitized media is not really what is important to your job unless you are working with lots of non-timecoded material. What you really need to re-create your job is the project folder on the internal hard drive. Take all of it, not just the bins or the project icon. Most short-term projects should have a project folder that can fit onto a high-density floppy. If you are working on "History of the World" or need to back up lots of graphics and music tracks from CDs, however, you need some other choices.

The first choice is a piece of software that ships with every Macintosh system called Compact Pro and every NT system called Stuffit. Don't use WinZip or it will corrupt your bins! These compression programs allow you to save your project folder in less space because of the removal of redundant information (Huffman encoding). Always be sure to save a file as a self-extracting archive so that you can open the compressed files on any system that will read the floppies.

If the backup process is unpleasant, you won't do it, so check the file size before you begin. You can find the size of a folder by clicking on it and then using Command-I to Get Info on the Mac or by right-clicking on the folder to get Properties on NT. Compression programs allow you to segment any file into smaller pieces and save it across multiple floppy disks. Some people are nervous about compacting files across multiple disks and for a good reason; if one of the floppies fails or grows legs, then the entire archive—your project—is unreadable. I have spoken to engineers who don't trust floppy disks at all, for anything. Besides, this method is tedious and slow if there is a lot to back up (more than two or three floppies worth or around 3 MB).

The best answer for backing up projects with graphics and CD music is some form of removable disk like a Zip or Jaz drive from Iomega. Now you have 100 MB or more of storage on a single cartridge for approximately US$100 for the machine and US$10 for the cartridge. Attach this to the native SCSI port on your computer, the one that is not accelerated. If you have an external 3D DVE box (the Aladdin) on the Mac, you may have a SCSI conflict until you figure out the correct order and proper termination. The Aladdin 3D DVE needs to be the last in the SCSI chain because it is internally terminated and Zip drives are either 5 or 6 for a SCSI ID. Alternatively, you can send your project over a fast network to a system that contains a drive suitable for backup. On a particularly big job, I backed up to a Zip drive on an Ethernet server on the hour. Although we lost power almost every afternoon while editing on location, we never lost any data.

Backing up to another hard drive on your system or another computer at your facility may not be good enough. I have had both the Boston Fire Department and Mother Nature ruin two different suites where I was working. (To be fair, the fire department was trying to save an historic building.) A particularly successful film assistant I know in Los Angeles makes two backups of the project as the film gets close to picture lock. He takes one, his assistant takes one, and they take separate routes home. It is Los Angeles after all. Avid technical support has a category in their database to record reasons for equipment failure. Earthquake is one of them. And having a floppy in your pocket also means you are more likely to get paid as a freelancer.

NON-TIMECODED MATERIAL

In the rush to complete a project, people throw anything and everything into their project just to get it done. Scratch audio is recorded straight to the Timeline, VHS material is cued up by hand, and CDs are played directly into the Avid without any thought about re-creating the job or, worse, starting over again if disaster were to strike. If the time is available, then seriously consider dubbing all non-timecoded material to a timecoded, high-quality format source. The potential slight loss of quality involved in dubbing will be the difference between quickly re-creating what you have done or matching things by eye and ear. If you cannot timecode all your sources, then seriously consider the DLT for the evening after you have finished digitizing all your media. Remember, a digital nonlinear project is never done, they just run out of time, money, or both.

After saying all this, I will mention a very useful technique for getting CD music into the Avid without actually needing a CD player. This technique is not nearly so dangerous as it once was since the introduction of Batch Import. If you have a CD-ROM player, you can import the CD audio directly into the Avid with one set-up step. If you have an older system, you might want to do this only if the rest of your project is at 44.1 kHz. Later versions (2.x/7.x on the Mac and later) perform a better up-sampling during import so that 44.1 kHz most of the time will sound fine at 48 kHz. Some early versions might bring certain tracks in a little too hot, so always check the audio quality for distortion before you use a track imported electronically from a CD.

Here's how you do it on the Macintosh:

- Open SimpleText or MoviePlayer. These programs ship with every Macintosh.
- Within both of these programs, open the desired CD and the track you need and choose Convert.
- Go under Options and set to: 16 bit, stereo, and 44.1 kHz.
- Save the file somewhere on a media disk and give it a name.
- Import as an AIFF file into your project.

All of these settings must be set right or the file will sound terribly distorted and unusable. You may have multiple copies of these programs on your internal drive and open a different copy the next time!

On Windows NT, you will have to get a type of program called a CD Ripper. These are small and easily downloadable from the Internet since they are favorites of people who like to make their own audio CDs with a CD burner. In the Goodies folder (installed from the Avid software CD-ROM), you will find a copy of WinDac32, which will do the job just fine. Here's how you do it on Windows NT:

- Open WinDac32 from inside the Goodies Folder.
- Unzip the file WDAC141.ZIP using Aladdin Expander (also in the Goodies Folder) if this is the first time you have used it.
- From within WinDac32 open the track on the CD.
- Save it onto a media drive as a WAV file and give it a name.
- Import it as a WAV file into your project.

You can now use the CD-quality audio without ever having converted to analog. This will save you some tape hiss and preserve the original dynamic range while eliminating the need for a tape deck to get the music or sound effects into the system; however, two weeks from now, when you need to redigitize the project for unexpected changes, you will have to have a way to track back to the original files. If you are using a version that has Batch Import, then you can point the Import dialog to the original file converted from the CD (or convert it again if you must). The Batch Import process will place the audio file frame accurately in all of the sequences that used it. You may have to use the Relink command if some of the sequences are hidden away in a closed bin or an archive, but this is a one-step process. If you have an older version of the software without Batch Import, then I highly recommend dubbing all audio to timecoded tape. You could also scrupulously back up the digitized or imported audio files (the file after it has been brought into the system and converted to an encapsulated OMFI file)

CONCLUSION

Beginners don't grasp these techniques from the entry-level course. They are usually a combination of changing some of your work methods and taking advantage of some unfamiliar functions. There are many other keys, modifiers, and tips and techniques in the Avid editing systems, but these are the main areas to concentrate on. Use the keyboard more and spend more creative time using trim. Start to think nonlinearly about navigation and the structure of the sequence. You will find your speed incrementing in leaps and bounds.

4

Administration of the Avid

Some people are lucky enough to have their Avid system administered by someone else. Hand them this chapter and go back to cutting, however, the day-to-day reality is that most people are responsible, at some level, for their own system. After all, if you have created a difficult situation for yourself, it is you that cannot go home until it is resolved. A smooth running session makes you look good, period.

This chapter explores the peripheral issues of owning and maintaining a professional editing system. This includes environment, media management, and networks. For the Avid administrator or post-production supervisor, getting media in and out of the system is as important as anything you do with it while editing.

ROOM DESIGN

It is easier to administer a system that is set up correctly and put into a room that is well designed for the Avid. Fewer constraints are placed on the design of a digital, nonlinear suite than a traditional, tape-based suite, but this doesn't mean you can just plunk the equipment down in a pile! There is flexibility to move all the equipment into another room and keep the bare minimum in the suite or move the noisiest parts like drives. There are technical implications to making cables longer, as you will see. Suites need to be designed so they can be serviced and simple maintenance can be done without disruption.

If you plan to use the suite as an online finishing facility, then you should treat it like an online suite. Use an external waveform and vector scope, high-quality speakers, a high-quality third monitor, and seriously consider a patch bay. There is nothing worse at 2 A.M. than carrying VHS decks around so that you can do two dubs. If connections are made through a professionally installed patch bay, there is a better chance the dubs will come out well!

Probably the biggest consideration after editor ergonomics is the noise level. Many people make the mistake of putting all their equipment in one room because the SCSI cables that ship with the system are so short. Longer SCSI cables are easy to purchase at your favorite computer or hobby store, but you should not use them for any reason. These will almost certainly cause weird problems and data errors later.

The slickest installations I have seen have completely moved the Macintosh and the drives to another room. They have extended the keyboard, or ADB (Apple Desktop Bus) or USB (Universal Serial Bus) cables, and the monitor cables. If you can also custom install a removable drive for projects, EDLs, etc., it will allow the editor to perform these loading and saving chores more efficiently. Truly, no devices have fans like the drives or CPU that cause much of the noise. Of course, if you have no central machine room and the editor must load tapes and digitize, there should be a connection for a video deck's video, audio, and control cables. If all the hardware is in another room, there should be telephone links between the two rooms (preferably with speakerphones for troubleshooting purposes).

The best facilities have gone back to the central machine room design after the initial years of isolated Avid suites. It is important to have a range of decks during the course of an edit, so multiple decks need to be available. Tying up a deck all day when it is not really needed is just as bad. A central machine room also makes it easier to get to the equipment for support or upgrading. There is nothing worse than working under a table with a flashlight in your mouth, the telephone in one hand, and a screwdriver in the other. Unfortunately, this happens way too often because of thoughtless room design.

If moving all the parts with fans is too elaborate, I have seen rooms where they used drive towers or JBODs (Just A Bunch of Drives). They were able to use the longer cables that come with a differential ATTO card for SCSI Media Share or Fibre Channel. The differential drives are no longer sold on the new system configurations. You can use Fibre Channel to put the drives farther away than just a foot or so (the length of the SCSI cables that come with the system). The use of differential SCSI led to one support call where the freelancer insisted that they didn't need drives because "they just plugged into the wall." Show your freelancers the layout before they start the night shift!

Many who build rooms with extended ADB cables use the Gefen Systems ADB Extender. This device amplifies the ADB signal so that it can be sent as far as several hundred feet. Monitor cables for video and audio can be extended quite a distance using standard audio and video cables but must eventually be boosted with distribution amplifiers (DAs). The new USB devices can be daisy-chained and connected to a hub to make them more flexible than the old ADB.

One of the nicest things about many of the Avid suites I have worked in is the addition of windows. There is nothing better to clear your head than to just stick it out a window for a few minutes, and having natural light is a nice change; however, I must fall back on the admonition that if you are doing color correction or shot matching of any kind, you need to have complete control of the lighting. Color temperature of sunlight changes during the day so that the shots digitized or color corrected in the morning may not look like those adjusted at noon unless the ambient lighting is indirect and consistent. If you have windows, make sure you can pull the room-darkening shades and keep glare off the monitors.

If you are also planning to do final sound mixing, make sure the room has been deadened. Applying sound-absorbing foam around the room can do this, or just make sure that the room has enough carpeting and wall hangings. Mixing in an empty office is probably a bad idea. Investing in good speakers and a real amplifier pays off quite quickly. You may also want to consider cheap speakers with an A/B switch to the reference monitors or pumping the Avid audio output through a standard home television monitor in the suite. There is nothing worse than listening to your wonderful sound mix at home and having it sound muddy from overpowered bass.

Personally, I feel no suite is complete without two phones, a trash can, a box of tissues, and a dictionary. You'd be surprised where people set up these temporary suites: attics, basements, bedrooms, storage closets, hotel rooms, boats, bank vaults, and ski lodges. Try to minimize any environmental impact like heat, dust, or jarring motion (like editing in the back of a moving truck).

ELECTRICAL POWER

The final and probably most important piece of equipment you need under any conditions is an uninterruptible power source (UPS). If you are running a system now without a UPS, you should nonchalantly put down this book and run to the phone to order one now. You are living on borrowed time. A UPS regulates your power, giving you more when your electric company browns out or less when you get a spike. If power fails completely, a UPS gives you enough battery time to shut down the system in an orderly fashion and avoid crashing, losing work, and potentially corrupting important media or sequences. A UPS makes your equipment run with fewer problems, and you will be able to charge for more productive hours of use.

The real question is not do you have one, but how much of one do you need? They are figured in the confusing scale of volt-amps. The math is not so hard if you can find out what each piece of equipment needs for electrical power for either volts or amps. The numbers are usually in the manuals or on the equipment itself. There is information on the Avid web site under the Customer Support Knowledge Center that lists the power requirements of each piece of equipment. But don't neglect connecting the tape deck. What happens to your camera original if the power goes out when you are rewinding?

Here's how to figure the size of an uninterruptible power source that is sold in volt-amp models:

$$volts \ x \ amps \ x \ power \ factor = volt\text{-}amps \ x \ power \ factor = watts$$

The power factor for computers is between .6 and .7. So you can look at the same equation as:

$$Watts \; x \; 1.4 = volt\text{-}amps \; for \; computers$$

watts for a Media Composer 9000 with 4 iS 9 drives, a PVW 2800, and a DLT = 1111 watts, so

$$1111 \; x \; 1.4 = 1555.4$$

And because you don't really trust manufacturer specs for the UPS and you buy more than you need to accommodate future expansion, you add another 33 percent on top:

$$1555.4 \; x \; 1.33 = 2068.7$$

You purchase the 2200-volt-amp model and sleep well at night.

Keep in mind that before you get a true blackout, you will probably suffer from sags and brownouts. These may cause the UPS to use up some of its battery power to keep you going until the Big One hits. Then, when the power comes back on, you can count on a serious power spike. A spike can cause damage to boards, RAM, and drives, and that damage may not show up until days, weeks, or months later as the parts start to fail prematurely. The fact that all power is going through a series of batteries and power conditioners with a UPS before it gets to your delicate equipment should give you a warm feeling in your stomach.

If there is any question about what to put on the UPS, imagine using that device full tilt when the power goes out. Ever see a one-inch machine lose power while rewinding a finished master tape? Not good. A Betacam SP will almost certainly crease the source tape if it is rewinding when the power goes out.

Even if all the lights and your monitors go out, you can always save and shut down quickly using just the keyboard. In an emergency, remember:

- Ctl or Command-9 to activate the Project Window. You want to save the whole project, not just the active bin.
- Ctl or Command-S to actually save everything.
- Ctrl or Command-Q to quit the application in an orderly way.
- Enter to confirm that you really do want to quit.

This sequence of keystrokes avoids the chance of corrupting the project from being shut down improperly and can be performed (if you really have to) with the monitors blacked out. A UPS has saved me literally a dozen times. After your first serious power hit, what is the real cost of replacing your system one board at a time?

ERGONOMICS

Human ergonomics has been written about at length in other places, so just a quick word about it here. Don't scrimp on chairs. They make the difference between happy editors and editors in pain. Get chairs that can adjust armrests, back, and height. Many people swear by armrests, and with a keyboard and wrist rest at about the same level, there is less chance of wrist strain. Keyboards can be put on sliding shelves below the workspace. Keep the back of the hand parallel to the forearm to reduce wrist strain. The relationship of chair, keyboard, and monitor cannot be underestimated as important to the creative process. Some editors, like multi-Oscar-winning editor Walter Murch, edit on the Avid standing up!

MEDIA MANAGEMENT

Let's discuss the nuts and bolts, the bits and bytes, of what happens when you put media on your system. The Avid editing application is an object-oriented program, which means that many things you do create an object. Digitizing media, rendering, importing, and creating sequences and bins all create different kinds of objects. It is the relationship between those objects that allows you to combine things in such interesting ways.

The limitations of desktop computers become apparent when you are dealing with hundreds of thousands of objects. Achieving object counts of this scale is not uncommon, and once the system begins tracking around 150,000 objects, you notice a slowdown. Features do not respond as fast, and you find yourself clicking and waiting more often. Buying more RAM can alleviate much of this. Because feature films require lots of footage to be online at the same time, you may want to add more RAM to your Film Composer. The more media you have, the more objects you have. If you have several 180 GB media drive towers connected to your system and they are full, you may easily have over 500,000 objects. If you are working on projects that require a lot of effects, you are rendering often and creating even more objects. Cleaning off the unused rendered effects is essential to making your system run at speed. In addition, if you experience slow response time, don't overlook just restarting — it defragments your RAM.

The number of objects can creep upward if you are not paying close attention, or if there are multiple shifts digitizing and they are not communicating. There is an easy way to find out how many objects are on the system right now. Go to the Project window, click on the Info button, and click on Memory (or choose it in the Fast menu at the bottom of the window). There is a display of the amount of objects the system sees.

Figure 4.1   Number of Objects in the System

The good news is that there are ways to reduce the number of objects without actually deleting material. Deleting is quick and easy, but if the material has no timecode, you are forced to either re-create it later or edit it back in by eye. Being able to clear off drives when they become too full with large amounts of small objects is a good argument for dubbing as much as possible to timecoded sources, but that takes time, too. As an administrator, you need to decide about policies regarding the speed and ease of replacing material and when it is appropriate to clean off drives completely. As you will see later, you may end up archiving certain crucial elements and redigitizing the rest.

The editing system only sees the media that is on the media drives in the folder named OMFI MediaFiles, 6.xMediaFiles, or 5.xMediaFiles, depending on the version of software you are using. This folder is created automatically and named by the software when the application is first launched and the drive is used for digitizing. The OMFI MediaFiles folder must keep this precise spelling and it must stay on the highest level of the file hierarchy. In other words, you cannot put this folder inside another folder and you cannot rename it. If you do, the media inside the folder goes "offline" and is no longer accessible to you when you try to edit. Also, any folder that is put inside the OMFI MediaFiles folder appears to be offline as well.

THE MEDIA DATABASE

Pre-Versions 7.x/2.x

In versions previous to 7.0/2.0, the way the editing system keeps track of all the media on all your drives is through a small file called the Media Database. With later versions on the Macintosh and all NT systems there is a slightly different set of rules that we will cover later. The Media Database is inside every MediaFiles folder and is continually updated every time something is added or subtracted to the media on the drives. If you change anything in the MediaFiles folder, the next time you go back into the software, the system stops and rebuilds the Media Database for you. This saves a lot of confusion and automates a very tedious task, but it goes to show how important that little file is. If your drives are too full for the Media Database to expand, for instance, you are in serious trouble.

The Media Database may not be able to completely rewrite itself because there is not enough space, and it will continue to interrupt you until it gets that space. If you are interrupted in the middle of normal editing by the Media Database being rebuilt, then you have a serious storage problem and should instantly clear space from your fullest drives. It is normal to see the rebuilding message, however, whenever you have done something on the Finder level to the media files, like copying or deleting, and now have switched back into the editing software.

Occasionally, as with all computer files that are written to and read from on a regular basis at high speeds, there is the chance of file corruption. This may happen if the drive doesn't update the Media Database absolutely perfectly because of a small drive malfunction. Later, when it comes time to read the Media Database, there may be problems. The information may not have updated correctly or it may be unreadable.

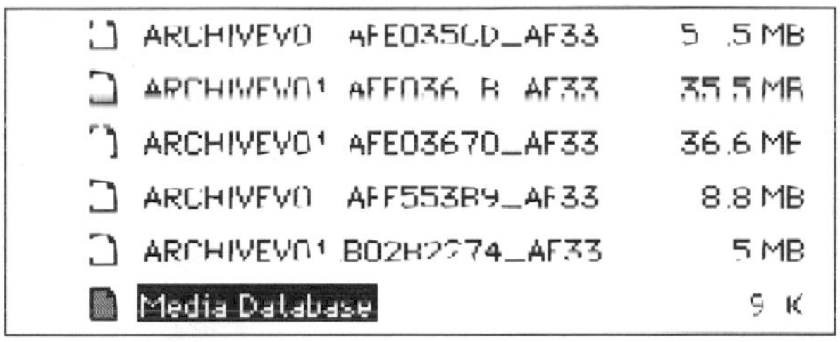

☐	ARCHIVEVO	4FE035CD_AF33	5 .5 MB
☐	ARCHIVEV01	AFF036 R AF33	35 5 MR
☐	ARCHIVEVO1	AFE03670_AF33	36.6 ME
☐	ARCHIVEVO	AFF553RY_AF33	8.8 MB
☐	ARCHIVEVO1	B02H2574_AF33	5 MB
■	Media Database		9 K

Figure 4.2 Pre-2.x/7.x Media Database

REBUILDING THE MEDIA DATABASE

There are two simple procedures you can choose from if you find that you have lost media that you know should be on the drives. You may suspect there is something wrong with the Media Database. You can go to the MediaFiles folder and find the Media Database by just clicking somewhere inside the folder and typing "M." You can also use the Find File command to find them all at once. Then, take the Media Database and drag it out of the folder and back in again. This may seem futile, but it is actually telling the editing software that something has changed and forces the database to rebuild. If this doesn't work, then you may have to actually throw away the Media Database by dragging it to the trash. Actually trashing the Media Database is a last resort and causes the software to start completely from scratch in making a new Media Database the next time you launch the application. This technique reinforces how important the information is in this little file, since this simple action may cause offline media to become online once again.

The Media Database only knows about the files in the MediaFiles folder and so only counts these files as objects. Anything dragged into another folder or out to the desktop is considered offline. This is an opportunity to manage your media in clever ways because it allows you to keep the media on the drives and still lower the number of objects. There is a wonderful third-party application written by Tony Black called MediaMover (download a demo from www.ran-domvideo.com). Avid no longer ships it when you buy a DLT, but you can buy it directly from him over the Web. What makes MediaMover so fantastic is that it lets you quickly and easily organize your media by projects.

MediaMover goes out and searches your media drives and returns a list of the projects it has found. You can then choose to move those media files out of the common MediaFiles folder and into a folder with the name of the project. You can also sort based on whether a file has been locked by the editor inside the Avid editing software. This makes it possible, for instance, to sort out only the non-timecoded media or all the graphic media in a project.

So when a project is done for the night, the night assistant or the post-production manager can take all the media that was generated that day and move it to its own folder. Clearly labeled, no one should mess with it by accident. Thus the number of objects is reduced, the system response speeds up, and the drives are simplified for the next project to start in the evening shift.

The next morning rolls around and it is time to start with the day job again. The night assistants, before they leave for the morning, go back to MediaMover. They move the evening project to its own folder and, employing the alternate function of MediaMover, they move back the day project into the MediaFiles folder. Simple, neat, and predictable.

Meridien-based systems, both Mac and NT, require the new version of MediaMover called MediaMover Meridien. Mac users on version 2.x/7.x need MediaMover OMF and 1.x/6.x users need MediaMover MFM. These different versions parallel Avid's legacy of changes in media files types.

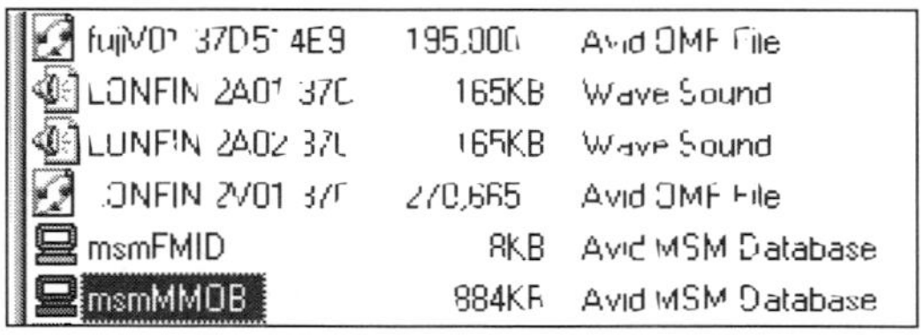

Figure 4.3   Latest Media Databases in the OMFI MediaFiles Folder

Media Files and Versions 7.x/2.x

The Media Database is replaced in versions 7.x/2.x Macintosh with two other files called msmOMFI.mdb and msmMac.pmr. The MSM type of media file in these later versions is designed to work on large media servers and to handle media in group user situations where multiple editors are sharing media. An important change for media-sharing situations is that with MSM, when media is changed in the OMFI MediaFiles folder, the editing on all the systems does not stop to update. Imagine being interrupted every time someone moved something around on one of your media drives if they were accessing it from another system! In order for the system to look at what is actually in the OMFI MediaFiles folder, you must choose Update Media Database from the File menu or open the Media Tool. After this, all offline media that is really in the folder is linked. This chapter will discuss relinking in more detail later. In later versions, relinking is more important than ever as a strategic technique.

COMPRESSION, COMPLEXITY, AND STORAGE ESTIMATES

None of this clever manipulation of objects solves the basic problem of running out of space on the media drives. It is only a matter of time before this problem occurs, and you should prepare for it in an organized way. There are several ways to tell when you are going to run out of space. The Hardware Tool under the Tools menu gives you a bar graph of how full the drives are relative to each other and percentages full if you have Tool Tips turned on. This is good for figuring out where to start digitizing the next job, but it does not give you the specific numbers to make you feel confident about the whole thing. Prior to versions 7.x/2.x, you could not use the Hardware Tool while in the Digitize Mode because the Digitize Mode marked off all empty space on the digitizing target drive. According to the earlier versions, this space is reserved just for digitizing and appears to be already filled when using the Hardware Tool during digitizing. In versions 7.x/2.x and all NT systems, the Hardware Tool has been changed and is accurate no matter what mode you are using.

If you need precise numbers, then open the Digitize Tool and choose the tracks you think will be needed the most. If you are working with material that has sync audio, then turn on all the tracks. But if you are working with mostly MOS (silent) film transfer that will be cut to a premade soundtrack, then get the estimate with only the video track turned on. Video takes a massive amount of space to store, even compressed, compared to audio files. The Digitize Tool gives you an estimate of how much space you can use.

With early versions, the estimates in the Digitize Tool do not show you how much space is on the drives. This is a common misunderstanding. If you think you have the space to digitize 30 minutes at AVR 75, you might actually try to do that, which would be a mistake. A file that big would exceed the Macintosh file size limit of two gigabytes for a single file. There is a certain amount of risk in having an important file be this large. If anything does happen to it, like corruption, you lose an awful lot of data all at once. Some animation or compositing programs allow you to render files larger than two gigabytes. The moment your project exceeds this size, it drops off the edge of the world never to be seen again. (The Avid software does not allow you to do this.) Thus the numbers in the older

Digitize Mode tell you not how much space you have, but the largest possible single file you can digitize and stay under the two-gigabyte file size limit. Also keep in mind on older systems, this means if you have a log shot that is digitized at a low AVR, it may be too large to redigitize at a higher AVR.

Versions 7.x/2.x and everything later (including NT) allow you to digitize one clip across multiple drives and across partitions of the same drives. In Digitize Settings, you can tell the system that when a clip is under a particular time length, say 30 minutes, and you need extra space, then jump to another drive and continue without a break. If the clip is longer than 30 minutes, it tells you that you do not have enough space just as it does now. Now by Alt-clicking or Option-clicking on the drive icon in the Digitize Mode you can toggle the display to tell you how much space is on the drive or how long your specific clip happens to be. Since you can now specify a group of drives, both the two-gig file size limit and the drive space of a single drive no longer restrict your digitizing.

If you select all the clips in a bin and choose Ctl-I on NT or Command-I on Mac (Get Info), this opens the Console window and gives you a total length of all your clips. This is a powerful way to see whether you have enough space on your drives to redigitize everything.

Unlike video or film, compressed images are judged by their complexity. A very complex image always takes up the exact same amount of videotape to record. When you digitize an image to disk, it is very important how complex a shot or image is because the more information and detail in frame, the more space it takes to store and the more difficult it is to play back. Playback from a disk-based system is a question of throughput, or how much information can be read from the hard drive, pushed through the connecting buses into the CPU and out to the monitors or tape decks. This is why a lower end system may not be fast enough because it cannot get the information from the drives fast enough to play all the information of a high-resolution image in real time with effects and audio.

Some systems actually drop frames, but Avid does not. Instead, you receive the dreaded video or audio underrun error message and playback stops. You need to decide how to get that section to play through, and this chapter discusses that later. It is remarkable that the Avid editing system will never drop frames during playback. This is crucial to trusting Digital Cuts as a visual reference to compare to EDLs and Cut Lists. Not many systems are always frame accurate for output.

All compression done by the Avid Broadcast Video Board (ABVB) uses something called Fixed Q tables. The data rate per frame digitized is based on Q tables (quantization look-up tables), which are created for each frame. Fixed Q means that a shot is compressed with constant quality, although the amount of compression may vary throughout the shot. A simple shot has few details and not many moving objects. As a result, it is compressed more because the human eye cannot perceive that the information is missing.

The Motion JPEG compression that Avid uses can figure out what the human eye can easily detect and so compresses less the images that require lots of detail. In general, M-JPEG compresses color more than detail, again because of the physical abilities of the human eye to see loss of detail before loss of subtle hue shadings. Even so, certain combinations of colors and subtle gradations do not hold up well. What this all means to you is that if your material is all talking heads, it will be compressed more and take up less space than a show that has lots of slow zooms of engravings, moving leaves, or crowds. This is why all space estimates on ABVB systems (and NuVista+) are given with high and low ranges.

The new Meridien video board found on Mac versions 2.5/8.0 and all NT systems has a different method of compression called JFIF, which is another variation of M-JPEG. This new compression scheme is based on the Rolling Q method for an image that may change in quality while keeping approximately the same amount of compression from shot to shot. This tends to ensure an overall higher data rate rather than an average, lower rate for Fixed Q. Although this high data rate will clearly improve the quality of complex images, it is also effective for improving the quality of simple images. Rolling Q does not try to compress a simple image more than a complex image; however, a simple image may not require that high amount of data and, rather than fill up the frame with noise or interpolated information, those images will use less data than the advertised fixed rate. This is why even with the new JFIF resolutions on the Meridien board, you may see different frame sizes with the same amount of compression.

What is the drawback to a fixed data rate? It takes more storage space. But with the plummeting price of storage making such financial decisions less important, it seems like a small price to pay. Always taking up the same amount of storage space allows the user to better estimate the total amount of drives required for a project. Expect uncompressed resolution to take up more than a gigabyte of storage for one minute of material (around 675 kilobytes per frame or kpf), and so expect Avid's new 2:1 resolution to take up about half that (300 kpf). This means that with a 90-gigabyte drive array you can expect to get about 76 minutes of uncompressed storage and about twice that with 2:1 compression.

There are three categories for digitized images: uncompressed, lossless compression, and lossy compression. Most compressed images on Avid systems are lossy, where redundant information is thrown away during digitizing. As you move closer and closer to uncompressed quality, you pay a higher cost for hardware and disk space. You must carry over every pixel of every frame, no matter how redundant that pixel is. Uncompressed images demand faster computers, wider bandwidth, and much more disk space on expensive striped drives.

Lossless compression is many times touted as better than uncompressed (or noncompressed, as some insist) because it takes less disk space. Lossless compression is more associated with programs like Aladdin's Stuffit or Compact Pro for compressing documents before posting them on the Web for downloading. The difference when compressing something variable, like a moving video shot, is that as the image gets more complex, the compression is less effective. Potentially, under a wide range of circumstances, a lossless compressed image could be larger than the equivalent uncompressed image (compression information and the less compressible image are added together). If the editing system is designed to take advantage of a low bandwidth as a benefit of smaller file sizes, you may have some playback problems. The system may impose a rollback, where a maximum frame size is imposed by throwing away information (lossy) when the frame size gets too big. If they don't do this, then they must prepare to handle even larger frame sizes than the uncompressed system and lose much of the benefit of lossless compression.

Only time will tell what technology eventually becomes the accepted quality, but the ironic situation at present is that although uncompressed is the goal for nonlinear editing, only two of the standard-definition digital tape formats use a component signal without compression. These formats, D1 and D5, are not for those on a low or even medium budget.

DELETION OF PRECOMPUTES

Predicting available space on media drives must go hand-in-hand with keeping track of rendered effects, or precomputes. Imported graphics and animation also take up space. Every time you render an effect, it creates a media file on the drive. Even though you may cause that effect to become unrendered or delete it from the sequence, that file still lurks on your media drive. This is actually for a very good reason, but it means you need to pay attention to how full a drive has become even though no one has digitized to it that day.

Deletion of precomputes is one of the most important things an Avid administrator can oversee. One of the most common calls to Customer Support is when, during a session, a system grinds to a halt because it is too full of thousands of tiny rendered effects. Are all those effects necessary? Most probably not, and now the editor or the assistant must be walked through the process of deciding what can go. One of your most important responsibilities in making sure that sessions start and end smoothly is to keep an eye on how many precomputes are on the system and how many are really important.

The Avid editing systems do not keep track of how many sequences are created during a project. There could be multiple CPUs all accessing the same media or archived sequences that are modified on another system and brought back. Since the system is so flexible, there is no way the system could definitively know the number of sequences created. What if the system were to make very important decisions for you, like deleting "unneeded" rendered effects? What would happen if you called up a sequence you had spent hours rendering to find that the software had neatly deleted that media automatically, thinking you were done with it? Just because you deleted the effects sequence from sequence version 15 doesn't mean you don't want effects on versions 1 through 14! There are too many variables, and this decision is too important to leave up to an automated function at this stage in the technology.

That being said, there is in fact some auto-deletion of precomputes going on under your very nose! But, as it should be with all automatic functions that cause you to lose things, it is very conservative and you may not even notice. The only auto-deletion of precomputes occurs when you are making creative decisions quickly and removing or changing effects. If you are rendering the effects one at a time and then quickly deleting them, there is at no time any opportunity for the system to save the sequence with those effects in place. There is no record that they will be needed in the future because they have not been saved. Saving happens automatically at regular intervals and, when you render effects in a bunch, they are saved as the last step before allowing you to continue. Every time you close a bin, you also force a save of the contents. That is why the software only auto-deletes precomputes when you are rendering one at a time and quickly removing or unrendering them by changing and tweaking. If a save occurs while an effect is in a sequence, then the precompute is not deleted automatically.

Every little bit helps to keep the drives unclogged, but you still must evaluate the amount of precomputes and delete them on a regular basis. This is really not so hard even though it is a little intimidating at first because it involves deleting material that someone (probably the editor) may need if you get it wrong. This chapter will deal with the isolation of precomputes when we look at efficient deleting strategies.

THE IMPORTANCE OF EMPTY SPACE

Remember the Media Database? Any file that keeps track of all the media is a concern if you fill your drives too full. That file must be allowed to enlarge to deal with the many files you add during the course of editing. If there is just not enough room, you may have media file corruption and eventually drive failure. A good rule of thumb is to leave at least 50 MB of each partition empty. It is also a good idea to erase media drives completely after a job has been completed. Don't get erasing a drive confused with reformatting! That is only a very last resort to save a dying drive.

BETWEEN SESSIONS

If there are multiple shifts all using the same system, then the changeover period may be as simple as shutting down, turning some drives off, turning others on, and booting back up. You want to minimize the actual disconnecting and reconnecting of SCSI cables (they are very fragile and subject to data errors if twisted). This is a good reason to change to MediaDocks or a shared media solution like Unity MediaNet.

If the media is moving to another system, consider putting the project files on the removable media. The project can be picked up and continued wherever the drives go. If these removable drives are too expensive, it may be time to begin restoring and archiving projects on DLT.

There will come a time when a project must be moved from one system or facility to another. It may be important to keep only part of a job and free up space for something new. These two situations require a better understanding of the sophisticated media management tools of the Avid editing systems and a few more steps on a regular basis to keep things running smoothly.

CONSOLIDATE

The best tool you have for moving media while inside the software is Consolidate. Consolidate can be used for two main purposes: moving media from multiple drives to one drive and eliminating unneeded material. You take whatever you need to move, either an entire bin of media or a finished sequence, and consolidate it to another drive.

There are some choices when you consolidate that may make things less confusing. First, you need enough space on your drives to double the amount of material you wish to consolidate. This is because the initial stage of Consolidate

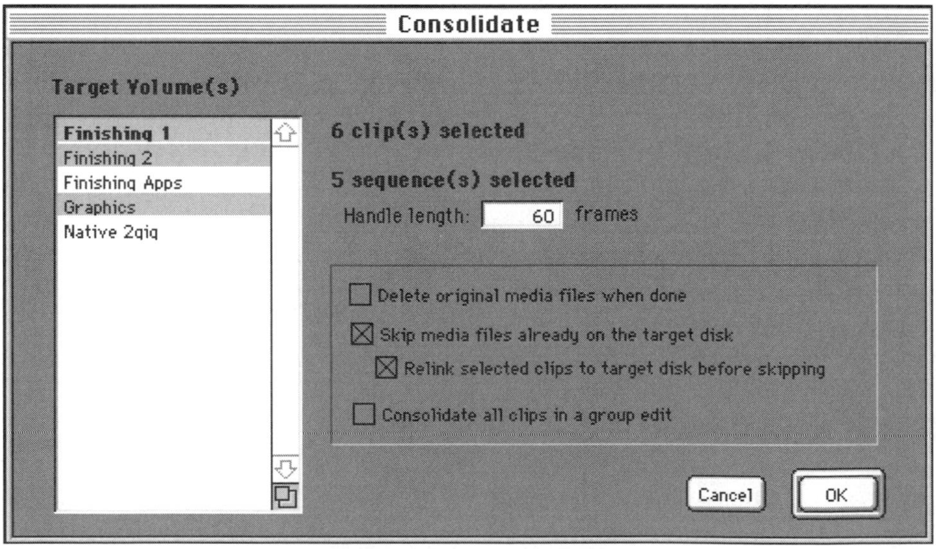

Figure 4.4 Consolidate

is to copy. Make sure there is enough space for your consolidated material to go to the new drive. You may need to break a sequence into parts if it is too long and at a high-quality resolution. Then consolidate each part to a separate drive. In recent versions, you are able to specify a number of drives for consolidation. Using the list of drives means that, even if you run out of space on one drive, the next drive in line will take the overflow material. The next drive will take the material until it is almost filled up and so on until the sequence is finished consolidating.

These are the two major reasons to use the Consolidate function, although as a quick troubleshooting tip you may choose to consolidate a clip that is not playing back correctly. If the clip plays back better after consolidating to another drive or partition, you may have a drive problem or you may have digitized the clip to a drive that was too slow to play back that resolution.

Consolidating a Sequence

The beauty of Consolidate is the advanced way that it looks at everything that is needed in a sequence and copies only that. There are, after all, other ways to copy media, but it is very difficult to tell at the operating system level what clips are really necessary to play a sequence. The Consolidate function will search all your drives for you, gather up only the bits you need, and then copy them to the desired drive.

Consolidating will break the sequence into new individual master clips and copy just the material required for the sequence to play. This creates shorter versions of original master clips because you are copying only the bit that is needed. You can then selectively delete unused media.

Consolidate is especially important at the end of projects when the final sequence has been completed and it is time to back up. Instead of backing up all the media, you consolidate first and back up only the amount of media that was actually used. It is also useful before sending audio to be sweetened on a digital audio workstation. Delete the video track from a copy of the finished sequence. Then consolidate this sequence and make sure the audio media files go to a removable drive.

So you select the sequence and consolidate it, and the system looks at the original master clips to determine exactly what is necessary. If you have an original master clip that is five minutes long, but you only used ten seconds of it, then only that ten seconds will be copied. The new ten-second-long master clip will have the original name and a ".new.01" if it is the first time in the sequence that shot is used. If you use another five seconds of the same original master clip, then you will have a second consolidated clip with ".new.02" and so forth. The Consolidate process also allows you to specify handles, or a little extra at the beginning and end of the clip, so you can make some little trims or add a short dissolve later. If the handles are too long, you will throw off your space calculations, so be careful; however, if the handles cause two clips to overlap material, then Consolidate will combine the two clips into one new clip rather than copy the media twice.

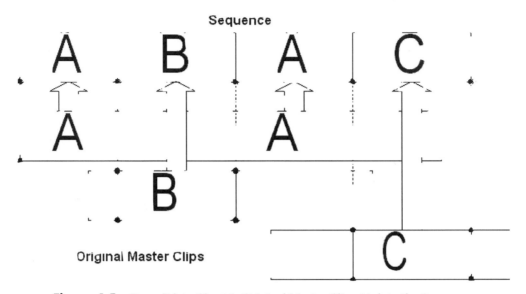

Figure 4.5   Consolidate Chart 1: Original Master Clips Link to the Sequence

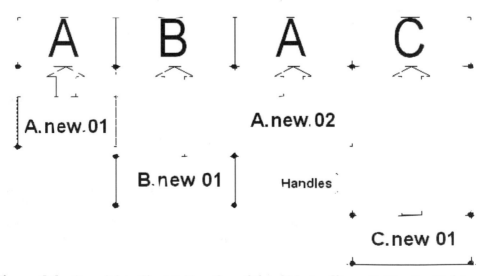

Figure 4.6   Consolidate Chart 2: New Consolidated Master Clips with Handles Link to the Original Sequence

Consolidating Master Clips

If you want to move media from multiple drives to a single drive, you can consolidate master clips. This will take all of the clips in a bin and copy them in their entirety to another drive. Notice the difference between "move" and "copy." You are really just copying and then must decide whether or not to delete the original. The copy and then the delete are the two steps that make up "move." Consolidating master clips is fantastic for being able to clear off multiple drives and put everything from one project onto a single drive or onto a series of drives so you can remove them or back them up. Consolidating master clips does not shorten material. It is just convenient and takes a complicated media management task and makes it one step. Be careful not to overfill any one partition.

If the idea is to move media from multiple drives to one drive, then you must ask yourself, "Have I used this target drive before for digitizing this project?" If the answer is yes, then you may already have some of the material you need on the destination drive. You don't need to copy it twice! Be sure to check the option "Skip media files already on the target disk." You would only use this if you were consolidating master clips and not sequences. Using the media already on the drives is perfectly fine. When the Consolidate function finds the long, original master clip already on the target drive, it will leave it alone, untouched. This is the best setting for moving media from multiple drives to a single selected drive.

There is a secondary choice that becomes available to you only when you check "Skip media files already on the target disk," and that is the somewhat confusing "Relink selected clips to target disk before skipping." So this seems a little obtuse, but let's work it out. Consolidating usually creates new master clips while copying the media (clip.new.01, clip.new.02). If you choose not to copy

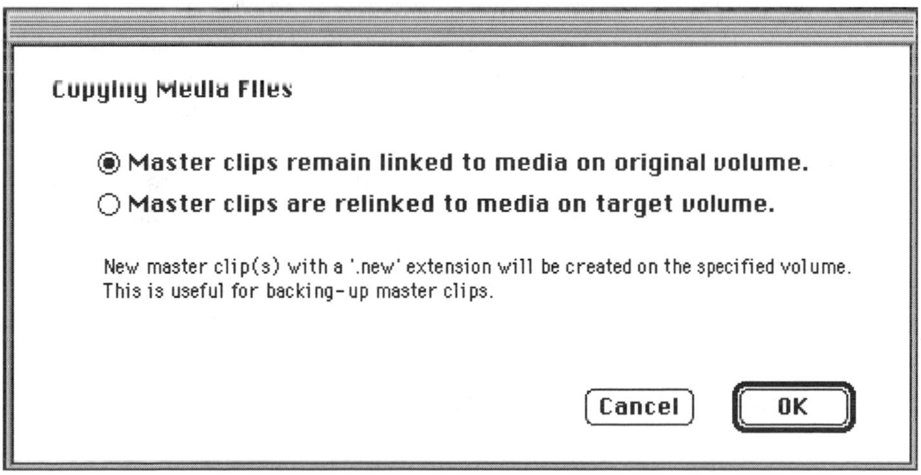

Copying Media Files

⦿ Master clips remain linked to media on original volume.
○ Master clips are relinked to media on target volume.

New master clip(s) with a '.new' extension will be created on the specified volume. This is useful for backing-up master clips.

[Cancel] [OK]

Figure 4.7 Copying Master Clips with Consolidate

media because it is already on the target disk (to skip the media files), then you would not want to create a new master clip—just use the old, perfectly good master clip. So what do you want to do? Do you create a new master clip and link it to media that is already on the target drive (clip.new.01) or take the master clip you have now and link it to the media on the drive? If you say yes to "Relink selected clips to target disk before skipping," then your original master clip will not get a new version with a ".new" appended; it will just link up to the media on the target drive and you are on your way. If you are moving the project so that it can be archived, or you don't really plan to work on this project in the near future, then say "yes" to this checkbox. You won't be creating a bunch of redundant master clips.

Why is this complicated choice so important? What if you are copying master clips to another drive so someone else can work with the media on another system. Then you can keep working on the same project on your system. If that is the case, then you don't want to link your master clip to the media that is going away, and you would not choose "Relink selected clips to target disk before skipping." When the consolidated drive is taken to the other system and your master clip is linked to the media on that drive, then what happens? Your existing master clip becomes "clip.old" and the consolidated master clip retains the original name. In other words, you don't end up with a ".new" on everything you will be using from now on. Instead, all the original media is connected to a master clip with an ".old" added, and the new, consolidated media is connected to the original master clip without anything appended to the name. You may choose this if you are planning to delete the old media anyway and so you will not end up, after a few consolidations, with clips that have ".new.new.new" appended.

Consolidate Summary

If you are consolidating sequences, choose the drives, the handle length, and say no to "Skip media already on the target drive." If you want to delete the original media after you have created the shorter consolidated clips, then check the "delete original media files when done." If you are feeling prudent, go back and delete the original media in the Media Tool after consolidation. If you are working with multi-cam, you should choose to consolidate all the clips in the group if you want to continue to have all the camera angles available in the consolidated sequence.

If you are consolidating master clips, then choose the drives and "Skip media already on the target drive." You most probably do not want to delete original media here and will proceed with more sophisticated media management later.

Fixing Digitizing Mistakes with Consolidate

Let's take the example of too many tracks. Perhaps you weren't paying attention when you were digitizing and digitized a video track to a voice over master clip.

You can take this master clip and make a subclip of the entire length, making sure that you turn off the tracks that you don't want. If you have digitized voice over and accidentally digitized (black) video, then before you subclip the master clip you should turn off the video track while it is in the Source window. The subclip will be audio only. Highlight that subclip in the bin and choose Consolidate under the Clip menu. The Consolidate function will copy only the media you want to keep. You will have a new subclip and a new master clip with only the audio tracks and can delete the original master clip to free up disk space.

You can also use this method to just shorten your master clips after you decide what part of them you really need for the project. In fact, many people like this method as a general strategy and digitize master clips that are quite long, maybe an entire scene or the full length of an already edited master tape. This creates fewer files on your drives for the computer to keep track of, fewer objects for an object-oriented program, and can speed up performance. The only drawback to this method is that if you try to redigitize the entire master clip at a high resolution, you may cause the media file size to exceed the two-gig limit. This is not a problem in recent versions that allow you to split large master clip media into pieces after a certain length and thus avoid the single-file size problem.

USING THE OPERATING SYSTEM FOR COPYING

If your goal is just to move an entire project to another drive, you are probably better off working at the desktop level. If you have been using MediaMover, your job is a snap; just copy the entire folder with the name of your project from each of the affected media drives. It is easy to do a Find File (Ctl-F on NT or Command-F on Mac) and copy every folder found with the project name. The situation may be complicated if there are two different resolutions to keep track of in this project. You may want to copy only the low-resolution material and leave the high resolution on the drives or vice versa. Planning helps here and, if on a Macintosh, I recommend that you use the Labels function provided by the Macintosh operating system. If you go to the Control Panel for labels, you can change what the colors represent. Change the blue label to Project X Low Res and the green label to Project X High Res. Select all media files immediately after you have digitized them and change their label color. You can sort the media files by label color and copy or delete only the files you want. This also allows you to track down those stray media files that always manage to escape even the most careful herding.

Whether you choose to consolidate or work on the Finder level, you need to keep track of all media. It must be searched for, backed up, copied, or deleted. The number of objects on the system should be periodically checked, and unneeded rendered effects or precomputes must be deleted on a routine basis. If your drives fill up, the session stops.

DECIDING WHAT TO DELETE

A place where an administrator or an assistant must be very careful and yet very efficient is the deletion of unneeded material. Sometimes this can be agreed upon mutually with the editor and you can eliminate all of the material for Show 1 when you are well under way with Show 2. Many times, different projects share the same material and you must be careful not to delete that which is needed by both. The Media Tool and MediaMover can both be used efficiently project by project, but this may not be good enough. You must find some other criteria to sort or sift by, protect certain shots, or change the project name of the material you want to keep.

USING CREATION DATE

Creation date becomes a very important criterion to look for individual shots, and it is often overlooked by many assistants. It does not work like the modified date on the desktop, which updates to reflect the last time someone opened and modified the file. The creation date is stamped on the clip when it is logged, so if the shot is logged and digitized on the same day, this becomes a useful heading to eliminate material that was digitized at the beginning of a project.

I especially like to use creation date as a heading in my sequence bin. I duplicate my sequence whenever I am at a major turning point or even if I am going to step away for lunch or dinner. When I duplicate the sequence, it time-stamps it so I know that I am working on the latest version. Then I can take the older sequence and put it away in an archived sequence bin (or several archive bins as they get too big). I have control over the exact times that I have stored a version, instead of leaving it up to the auto-save, and I can keep the amount of sequences in any open bin to a minimum. Sequences will be the largest files you will work with. In an effort to have the bins open and close quickly and not use up too much RAM, I try to keep the bins small and only have the latest version of any sequence.

If you are trying to determine which sequence is the latest version and the editor is not present, creation date is the best tool. There is always the chance that the editor duplicated the sequence and continued to work on the old one, but you should have an agreement with the editor about how to determine this crucial fact.

CUSTOM COLUMNS

There are many other criteria you can use for sorting and sifting. If you plan it well, you can create custom columns to give you an extra tool to work with. Some people will create a custom column with an "X" or some other marker to show whether a shot has been used. One way to use sifting powers of the bin is to change the sift criteria to "match exactly" and have it search for a blank space in a custom column. This way, any shot that has *not* been marked is called up. The Media Tool is the only way to search for media that is online across bins. In current versions, you can create a Media Management bin view, which can be used in the Media Tool. Your custom columns will show up there, too.

MEDIA RELATIVES

Another way to find out if a shot has been used within a project or sequence is to use the "Find Media Relatives" menu choice that is in every bin. This is the best way to search across bins to clear unneeded shots, other than consolidating and then deleting the old media.

The most useful way to use media relatives is with sequences.

- Open the Media Tool. You have a choice whenever you open up the Media Tool to show master clips, precomputes, or individual media files for all the projects on the drive or just this project. I like to show "All Projects" when checking for media relatives because many times the bins I am working with came from another project.
- Open all your relevant sequence bins for this method to be accurate.
- Put all the sequences that you want to work with into one bin.
- Highlight all the sequences and then choose Find Media Relatives from that sequence bin.

The system searches all open bins and the Media Tool, and highlights all the clips, subclips, precomputes, and media that you will need to keep.

- Go back to the Media Tool and choose Reverse Selection, another choice in the pull-down menu for that bin.
- After reversing the selection to highlight all the unused media hit Delete.

Keep in mind that Find Media Relatives does not unhighlight something, so before you begin this operation, make sure nothing in the Media Tool is already selected. Also, give everything selected for deletion one final look. This is the last time you will see these files, and you don't want to trust anything or anyone except yourself at this stage — there is no undo. Don't do this when you are tired! Wait until the morning and do it first thing if you can.

If you choose to show precomputes in the Media Tool, you see all the rendered effects for a project or on your drives. You can use the above method, Find Media Relatives, to track down the precomputes used for the sequences you've selected and then, remembering to reverse the selection, delete the unneeded

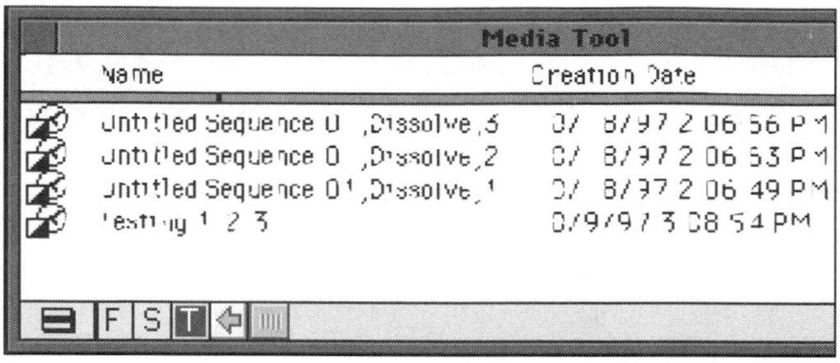

Figure 4.8 Showing Precomputes as Media Relatives

precomputes. This process can be done every day if you are working on intensive effects sequences, like a promotions department. It doesn't need to be done unless you need the space or memory requirements are climbing into the hundreds of thousands of objects. There is always the chance you might delete something you want to keep, so be careful.

Occasionally you will want to find exactly where a specific file is and for some reason, Consolidate, MediaMover, or the Media Tool are not sufficient to perform the task you need. In recent versions you can use a new function called "Reveal File," which will go to the desktop level and highlight the media file associated with a specific master clip.

If you have a system without this choice, another technique uses the Media Tool. In the Media Tool, choose to display the individual media files. You don't need to see these very often, but there is one very useful procedure.

- Display master clips and individual media files.
- Highlight the master clip.
- Choose Find Media Relatives.

The individual media files associated with that master clip are highlighted. There may be up to nine media files associated with any master clip: one video and eight audio tracks that were originally digitized with it.

- Choose the track you are interested in and click on the name, not the icon, of the individual media file.
- Use the copy function (Ctl-C or Command-C) to load the name into the clipboard.
- Go to the desktop and use the "Find File" (Ctl-F for NT or Command-F for Mac).
- Paste the long media file name into the Find File dialog box and search for it.

This may be the best way to discover duplicate files and eliminate them.

LOCK ITEMS IN BIN

If I have achieved my goal of making you slightly paranoid about deleting material during a project, then you will approach this task with the proper amount of stress. There is a great way to relieve some of this stress that does not require a doctor's prescription. A few years ago I spent a week at a major network broadcast facility observing their operations. It was pointed out that there was a need for someone who knew nothing about the individual projects being edited to come in late in the day and clear drive space for the next morning. On longer jobs, this person would be familiar with the exact needs of the editor. In this case, they were working on as many as five or more projects a day. At the end of the day, some material needed to stay—material that had been grabbed from their router and therefore had no timecode—but the majority of the material needed to be deleted. This is where the idea for "Lock Items in Bin" originated.

It was supposed to work like this: As the editors are gathering shots and using them, they decide that they need some shots for the continuing story tomorrow and clone the master clips (not the media). They Option-drag and clone the master clips into a bin named "Stock Footage." At intervals during the day, the editor goes to the stock footage bin and selects all (Ctl-A or Command-A). The editor chooses Lock Items in Bin from the Bin pull-down menu (or right mouse click in NT, which is the only backdoor to get this to work on an Xpress). A lock icon appears in the Lock heading that is displayed as part of the view for that bin. The file is automatically locked on the desktop level so that if some intrepid assistant decides to throw everything away, they see a warning that certain items are locked and cannot be thrown away. Any good assistant, however, knows that on the Mac if you hold down the Option key when emptying the trash that you can throw almost anything away and cheerfully ignore these pesky warnings. On NT you would need to go to the file, right click on the icon, and unlock the file under Properties. You can select many files simultaneously for locking or unlocking, but there is no shortcut to just dragging them all to the Recycle Bin. You can see whether a media file is locked on NT by showing Details in the Folder View and looking for an "R" under the Attributes column for Read Only. An administrator on an NT system can set permissions for certain users so they will not be able to delete media directly from the desktop.

A happy side effect of the Lock Items in Bin command is that it can also be applied to sequences. Subsequent copies of that sequence are also locked automatically when the sequence is duplicated. This helps avoid the all too common problem when people have mixed clips and sequences in the same bin and they decide to delete files. They select all and delete and then—hey, where did my sequence go? Certain types of objects are automatically selected as choices to be deleted when you select all in a bin. In the latest versions of the Avid software, the sequence is automatically not checked for deletion if it is mixed in a bin with clips. Select all with a bin of only sequences and they will indeed all be checked for deletion. Deleting a sequence by accident was usually someone's first visit to the Attic, only to recall the version of the sequence they worked on fifteen minutes ago!

There are ways to outsmart the Lock Items in Bin command if you are determined. You can duplicate the locked clip and then unlock the duplicate. Because both clips link to the same media, you can delete the media when you delete the second clip. You can also unlock the media at the desktop level and throw it away without even opening the editing application. But you can't delete the media until someone, somewhere, unlocks it. Since there are really no permissions required to operate the Macintosh Finder, as in the SGI or Windows NT, it is really the honor system for media management.

LOCKING BINS

A new, slightly subtle feature has just been added to NT releases 3.1/9.1/2.1, which is the ability to lock an entire bin. This becomes activated only when sharing projects over a Unity system. An editor who is busy at work on an important sequence or is organizing a large and complicated project doesn't want another editor to start changing things in the bin; however, the other editor may be working on the same material and need access to the latest version of the sequence or the latest subclips and custom headings.

So what is a good compromise? The editor who owns the bins can decide if anyone will get "write" access or "read-only" access. Write access allows anyone to add things to the bin or make changes to *anything* in the bin (like duplicating, deleting, or trimming sequences). No other editor can get write access to the bin while it is being used by another editor with write access. The write access is displayed by the green unlock icon at the bottom of the bin.

Read-only access is the limited ability to work on the bin without making any permanent changes. The read-only status is displayed by a red lock icon at the bottom of the bin. This means you can view clips and, if you really want to make changes, you must copy the clips or sequences to another new bin. Now you are forced to work on your own, separate version of anything. Any bin can be opened at anytime by Alt-clicking to open the bin.

Restricting all changes would unfairly hobble fast-working professionals. This would mean you could view but not mark point or add locators. Flexibility was added so you can make some minor changes to the read-only bin, but the changes will not be saved when the bin is closed. There is a warning when you close the bin (and at the first auto-save) that your work will not be saved. If you have made changes in the read-only bin, cancel the bin closing and drag your work to another bin. This flexibility is a trade-off with stupidity. This short-term capability is a convenience, but it should be clearly stated to any inexperienced user the full consequences of working in a read-only bin as a standard procedure. Sure, do something just to look at it quickly, but if you like the change you must drag the work to another bin.

Obviously, it is important to give write access only to qualified people who have the responsibility to own up to the changes they make. Ronald Reagan said, "Trust, but verify," so there is a log of any user who has "dirtied" a bin. When a user dirties a bin, the asterisk appears next to the bin name in the header bar across the top of the bin. This is a hint that there are unsaved changes (on the Mac this is a diamond icon). The log is accessed by right-clicking the mouse on the colored lock icon at the bottom of the bin.

Here is how to lock a bin from write access by other editors:

- Select the bin or all the bins in the Project window.
- Under the Clip window choose Lock Selected Bins.

The bin name will now have an asterisk next to the project name in the locking editor's bin. In all other editors' bins the bin name will appear in bold and the name of the person who has the write access will appear in their project window next to the bin name. If you want write access to that bin, then give the owner a call and ask permission for them to release their control.

CHANGING THE PROJECT NAME

There is another way to organize shots for easy media management: change a master clip's project name. There can be confusion about exactly which project is associated with which media because it is so easy to borrow clips from another project that you may not even know that certain shots are from another project unless you choose to show that information in your bin headings. This is why I create a custom bin view just for media management. It shows project name, lock status, creation date, tracks, video resolution, disk, and any other customized headings that relate to media management. If the clips have been borrowed from another project and brought into the new project, even if you redigitize them, they retain their connection to the project in which they were logged.

Project affiliation makes a big difference when it comes to redigitizing at a higher resolution. Because all the shots from one tape are digitized at once, if you are trying to redigitize from Tape 001 in Project X and the clip is really from Project Y, the software considers these two different tapes. While batch digitizing, the system will ask for Tape 001 twice since it really should be two separate tapes if logged correctly.

There are several steps to changing project affiliation. The first thing to keep in mind is, as mentioned in Chapter 2, project name is part of the tape name. The trick to changing the project name is to change the tape name to a tape name from the right project. In recent versions, all tapes have a Project column in the

Select Tapes window when assigning tape names. This allows you to see which project each tape is from.

In Capture Mode, whenever you show the list of tapes that have been digitized and you check "Show other projects' tapes," you might see multiple Tapes 001 from other projects. You occasionally need to use the Scan for Tapes button in recent versions to make sure the system has updated all the tapes digitized and used on the system.

Here's the best way to change the project name of a master clip with 2.5/8.0 Mac and 3.0/9.0/2.0 on NT:

- Open the new project.
- Open the Media Tool and show the original project media files.
- Consolidate the master clips into a new bin in the new project.
- Do not skip files that are already on the drive.

On older systems, here is the best way to change the project name:

- Open the new project.
- Open the bin from the old project that contains the shots that you want to permanently change to a new project name.
- Create a new bin and copy the clips into the new bin.
- Select all the clips.
- Go to the Modify command under the Clip menu and modify the source name.

If the tape name you want has not been used before in this project, then you can create a new tape name. You can choose an existing name if you made an earlier mistake. You might need to rename these clips so they all come from an existing tape in the new project. As you finish with the modification process, you get a series of warnings that are important if you are working with key numbers in a film project since the new clips need them re-entered after you modify the tape name. A film project will not allow you to change a source name of a clip without unlinking first. Otherwise, don't worry and just check OK.

Now the clips are associated with the new project as far as the Media Tool and the headings in the bin are concerned. The change has not really affected the actual media file at this stage, just the master clip in the project. This is not enough for MediaMover to recognize there has been a change since it only looks at the media file and not the project. The project information about a media file is actually recorded in the media object identification (MOB ID—it has nothing to do with gangsters). The MOB ID is attached to the media file when it is digitized, and the media file itself must now be changed to update this information. Here is another use for Consolidate: You need the space on the drives to copy the files you want to change.

Change the project name of the media files like this:

- Select them all and Consolidate with the following options checked:
  - Have the old name link to the new media.
  - Do not skip over the media if it is already on the drive.
  - Delete the old media when you are done.

If you want two sets of the media—one set of media in one project and one set in the other—do not delete the original media after you consolidate.

BACKING UP MEDIA

Now that you have eliminated the unneeded material by deleting, whether it is digitized media files or precomputes, you can now evaluate the choices available for backing up. As very large media drives become a reality, there may be a room full of gigantic backup drives in your future. Clearly, this is the fastest way to copy media and, if it is done over a very fast network like Fibre Channel, those drives could be anywhere in the building or, eventually, the world. Perhaps there is a future business for someone to set up thousands of drives in a warehouse somewhere in the desert and have people rent out the space for backing up over the Internet.

But that day is a few years away, and for now, you probably have to deal with digital linear tapes (DLTs) or another form of archival tape backup. Until recently, 8mm drives were considered along with DLTs, but the size of the 8mm capacities has not grown as fast as the size of the DLTs, although 8mm drives are cheaper. According to the advertisements for different kinds of backup products, it may seem that an 8mm drive is more than adequate for your storage needs. Maybe you can pocket the extra money and use it for that industrial cappuccino machine. This is because most advertisements feature the amount of storage they are capable of holding when the data is compressed. But you have already compressed. You are working with Motion JPEG and Avid's compression schemes that have already squeezed the heck out of it. These files are not getting any smaller, which means you should always choose to avoid any further compression as an option during your backup session. Further compression takes extra time and adds a layer of unneeded complexity to already very large and complex files. You may be courting data corruption, and that's not a cheap date!

The linear part of the DLT is what most people object to when using them, plus the fact that they can be rather slow to back up and just as slow to restore. Just like hard drives, they too get faster all the time. The fastest DLT that Avid sells backs up at the blinding speed of 5 megabytes per second, and it can back up or restore 18 gigabytes in an hour.

Many people use DLTs to archive material that has no timecode. If they were to delete and redigitize or reimport non-timecoded material, they would have to re-edit it in by eye. If the material is graphics, then they may have many layers to step in to re-edit and potentially some very complex sequences to modify. I still recommend dubbing non-timecoded material to a timecoded format, but sometimes, especially for fast turnaround projects or lots of graphics, this is just not practical. You also lose the quality benefit of keeping the images completely digital and uncompressed until they are imported, unless your timecoded medium is also uncompressed. Dubbing non-timecoded material to tape with timecode is still the simpler and more reliable way to get material back on the drives and get it to link to your sequence. If you are using the more recent versions that have batch import, then reimporting and relinking graphics and animations is now a one-step process.

Many times it is actually faster to redigitize a shot or two than rely on a slow and cumbersome restoration process, which is why people don't back up their media as much as their project information. The exceptions to this would include video that has been carefully color corrected during digitizing using an external color corrector and the color correction information has not been saved. Also, there are many instances where the original tapes become unavailable because either they are sent to another facility or, working without a net, they are recorded over. Again, the question is not if your hard drives die; it is when. What is truly important is not how easy it is to back up but how easy it is to restore.

The professionally paranoid know that something is really out to get them and you can never be too prepared. For people like this, it is usually a good idea to do a full backup of all material after the first day of digitizing. From that point on, incremental or normal backups can be scheduled every night to store only the material that has been added throughout the day. You may want to even consider backing up over a fast network so that you can back up all the editing systems from a central location. With the size of storage on even one machine easily overwhelming the capabilities of a smaller capacity DLT, this may be just wishful thinking. You may need a DLT for each of your important systems. Figure out how much it would cost you to lose that information for even four hours and then compare it to the cost of a DLT.

A new choice for backup software that has begun to ship with DLTs when you buy them from Avid is MezzoArchiver from Mezzo Technology. This software backs up in the background throughout the day. You can identify all the material that should be archived by dragging the project folder to the graphical interface of MezzoArchiver, and leave the DLT connected to work in the background. You will find at the end of the session that only the very last material digitized or rendered needs to be backed up before you can go home. MezzoArchiver also allows you to target specific clips for restoration instead of the entire project. Since MezzoArchiver can identify all the material attached to a project, search for it on all the drives connected, and back up incrementally, you are operating with a safety net. For an emergency, like when a single non-time-coded video file has been deleted, restoration will be a fast and painless process.

The only reason that a DLT is not a foolproof way to back up media files for archival purposes is that over time, there may be a significant change in the structure of media files. This may occur because of a software change, as in versions 7.x/2.x, when the structure for the media files changed from MSM to MFM to be compatible with NewsCutter and the EditCam and in the process became OMFI native. MSM ensures that all media files are "server ready" no matter which edit system you are using. If you upgrade to that version of the software, then all of your media files must either be converted to the new format or redigitized. When Avid made the switch from the NuVista+ video board to the Avid Broadcast Video Board, there was a change in the size of the frame as well as a jump in quality. You had to convert the NuVista+ media, which was digitized at 640×480 square pixels to the 720×486 non-square pixels. No pixel-based images are improved by expanding and distorting. The way to take advantage of the new technology was to fall back to the old technology and redigitize from tape. In the short run, or if you are not planning to upgrade any time soon, the DLT is your best answer. A DLT will allow you to keep entire projects, with all the rendered precomputes and non-timecoded material, all in one easy-to-restore package that takes up very little space on a shelf.

OTHER OPTIONS

Other options don't include backing up everything—just those items that can't be redigitized. The slower optical drives are not supported by Avid with the newest software for playing or digitizing—even the Avid Optical drives. Many people are considering the new removable one or two gigabyte cartridges like the Jaz Drive by Iomega because they are as fast as a slow hard drive. Some people are even brave enough to play back audio from these drives (not supported)! You can improve performance of these drives by buying two and striping them (really not supported). If you misplace one of the striped cartridges, the one you have is not readable.

The removable drives are far more expensive per gigabyte than the DLTs, but then you need to be choosier about which media actually gets backed up. You should put all non-timecoded material on one partition so that only that partition is copied to the backup drive.

BACKING UP PROJECTS AND USER SETTINGS

After all these methods of backing up media are considered, don't forget to back up your projects, too! This is the most important thing you back up and is all you need to re-create any project, except for non-timecoded material and rendering. You may want to back the project up to several places as well as putting it on the DLT.

There is the common question of backing up User Settings. You should be able to reproduce User Settings in about five minutes, but some people resist it like a trip to the dentist. There is no way to lock your User Settings to keep unauthorized folks from making a few "improvements" or just accidentally changing them, so it is a good idea to back them up somewhere safe and hidden. Many freelancers carry their User Settings on a floppy, and other, more intrepid instructors or field support technicians carry their own small, fast hard drive with everything they could possibly need.

When software versions change, especially substantial version changes, you should always remake your User Settings. I know this sounds tedious, but re-creating them solves many unusual and unpredictable problems, especially if your customized keyboard is complex. You may be trying to access menus that have been moved or deleted! Or there may be more subtle problems that don't immediately appear to relate to User Settings. One of the first things Customer Support asks you to do if you are getting unusual behavior, like a common feature suddenly not working, is to re-create your project, bins, and User Settings. If the software version has not changed recently, having a backup of your User Settings may be enough to fix the problem, rather than having to re-create them from scratch.

NETWORKS

Any good administrator must evaluate how a network can improve their operation. One of the smartest broadcast engineers I ever met told me he was studying networks at night school. Very much like plastics was the secret answer in the 1960s, networks may be the not-so-secret answer in the next decade. Networks, as they exist today, are slow, but just like hard drives and DLTs, they are getting faster all the time.

The question for you is: "How much is it worth to have a network?" and plan your investment accordingly. As you may suspect, the more money you invest, the faster you can go. Whatever you do, do not rely on the old standby Apple LocalTalk. This is generally incompatible with the Avid and has been known to cause some pretty serious problems during digitizing and digital cuts. Besides, who wants their sessions interrupted with e-mail notices? The only thing you might consider using LocalTalk for is tying into a printer, but you will have to restart the computer to activate it. Over time, this will prove to be more awkward than useful and eventually you must get a printer that does not need LocalTalk, uses a LocalTalk converter, or connects directly to the serial port of the Macintosh.

Better to consider Ethernet. It may not be the fastest network choice out there, but it sure is easy. All systems currently shipping from Avid are Ethernet ready.

Plug and play and all those printer problems go away if your printer is Ethernet ready, too. You still probably don't want that e-mail icon flashing in the right-hand corner, though. Any time you are asking your system to be interrupted by incoming signals, you are asking for some kind of error message during CPU-intensive tasks like digitizing, digital cut, or rendering.

Ethernet can be easily expanded by the addition of an Ethernet hub. For very little money you can tie all your editing and logging systems together on one local network. The best reason to use plain vanilla Ethernet, technically known as 10 Base-T, is to share project information and small graphics. It is also a good way to set yourself up to back up just the contents of the internal hard drives with project, bin, and script information every night. You can use it to send the occasional media file, but there are better options for that which we will cover later in this chapter.

The 10 Base-T name refers to the size of the wire connecting all the systems. Think of it as a thin, but adequate pipeline that maxes out at a total capacity of ten megabits per second. Don't expect to be blown away by the speed, but you will appreciate the simplicity. For serious network use, consider adding some more money and another circuit board to the equation. The next step up is Fast Ethernet with 100 Base-T and, as you have probably guessed, it is faster and uses a thicker cable for the connections. It also requires a small addition to your CPU on older models, so you might need an extra slot inside your computer for a circuit board that will allow you to accelerate the Ethernet. The more recent CPUs are shipping with 100 Base-T already installed. This allows you to consider larger graphics, digitized audio, and the occasional video media file sent over the network. Many people use this to tie into their digital audio workstation.

MEDIASHARE

SCSI MediaShare uses a special piece of hardware, a switching device that Avid supports for three editing systems to connect to the same drives at a discrete distance. Editor A can be cutting using material from the group of drive towers while the assistant is digitizing on another system to the same drive tower. Or, Editor B can be working on a second system while Editor C can be outputting a digital cut. There are strict limitations with SCSI MediaShare because it is still a SCSI system and not a new kind of network. SCSI MediaShare is limited to three drive towers and can reliably work with all single-field AVRs. Although many people push these limitations and can work beyond them, things don't work quite as smoothly. As of 7.0/2.0, SCSI MediaShare is no longer supported. If you have an older system, you will still need to deal with these limitations, but the newer systems use MediaShare Fibre Channel or Unity MediaNet, which we will discuss later.

There are also administration limitations to SCSI MediaShare because it relies on a piece of software called MS Scanner. Only one system, the administrator, can have the privileges to do media management on all the drives, and this one person can decide the privileges for the entire group. If you do not work at the administrative system, you cannot write (digitize, render, import, etc.) to certain drives unless you have permission. Changing administration privileges or performing drive maintenance requires shutting down all the systems. With lots of media online, several towers full of low-resolution clips, restarting can take around twenty minutes.

If you have a traditional feature film setup of one or two editors and an assistant and this setup will not change during the project, then SCSI MediaShare may be adequate. If you have a fast-changing situation, it may not give you enough flexibility. It will never be fast enough for online-quality material and will never go beyond the current capabilities.

FIBRE CHANNEL

A Fibre Channel-Arbitrated Loop is so fast that you will find yourself not copying files across this network, but actually using media files from a central group of Fibre Channel drives that everyone shares. In the past, this functionality has been available only when using MediaShare, but now Avid has released MediaShare FC. Fibre Channel makes it possible to access large media files at speeds that will equal or exceed the amount of time necessary to play these shots back in real time. Theoretical speeds of transfer of 100 MB per second will allow five simultaneous users at AVR 75 or ten users at AVR 8s. There is no server required, no IDs, over 100 devices per loop, and up to 30 meters between drives or workstations. Fibre Channel is the network to get if you are serious about sharing or exchanging media files.

AVIDNET

Fortunately, Avid has made it even easier and faster to use these kinds of networks by creating its own file transfer protocol called AvidNet. This means that Avid uses the wire connections but designs its own way of sending the information through the network. With AvidNet, you can send material only between Avid systems that use the AvidNet protocol. It allows transfer of material faster than standard Ethernet by making up for inherent compromises in the Ethernet protocols. AvidNet comes standard on some models and is an option on others. If you already have it installed, it may be the easiest way to send small amounts of data between systems and across platforms.

AvidNet is also easy to use since it works from inside the editing application. What could be easier than an inbox and an outbox? The user at the first station copies the material to be sent and puts it in an outbox on the system, then alerts system 2 that the media is ready. System 2 receives the media files automatically.

You can send media as an OMFI file, a sequence, or a sequence with the media attached. You should consolidate the sequence first, since when you send OMFI files attached to a sequence you send the entire original file, even if you have only used a small part of that file in the sequence. Several Macintosh versions of AvidNet actually consolidated the sequence while transferring. This is convenient, but slow. You may find that you get more consistent results overall if you consolidate the sequence first yourself and then transfer it and the media to another system. This is the best way to send just what is needed for another editor to continue to work on a sequence at another station.

SERVERS

Any administrator should also consider setting up a server. Simply, a server is a basic stand-alone computer fast enough to handle large transfers of information. It is connected as a central storage place for resources important to everyone working on the project. It may be a place to store the final version of any graphic, logo, or script, so the chances of someone using an older or unapproved version of something are reduced. Many times people end up using the wrong version of something because it was changed on one system and not copied to all the other systems. There should also be a good policy to enforce and maintain a level of compliance.

On one project, we were so concerned about individual editors changing the tape named bins that everyone was asked (repeatedly) to digitize and edit from bins that were on the server, but not on their system. In other words, they would access all bins over the network. The server icon would appear on their desktop like another drive, and they could open any bin stored there just as easily as any bin actually on their own system. If they wanted to customize or modify a bin, they were forced to copy the bin to their individual system. From that time forward, it was their own responsibility to keep track of that bin and name it for themselves. This maintained a pristine copy of every original bin in the project in a place where all editors could get it at any time. The downside was if the server crashed, they crashed, too. Unlikely but possible, and they would still have their own Attic.

A server is also a good place from which to base a facility-wide backup or just to connect a Zip drive to back up central bins every so often without disturbing anyone else. It is also a good place to connect a printer for everyone to use. An assistant can be printing out bins non-stop and never interrupt any editing or logging, although the server could conceivably be used as a logging station in a pinch. The server can be a good place to do database searches with a program like Retrieve-It! and MediaLog or to make EDLs with EDL Manager. You don't need a computer full of fancy circuit boards to open a bin! Besides, what else are you going to do with that old Power Quadra 950?

AVID UNITY MEDIANET

Avid has developed a new way to share media and metadata (projects, bins, sequences, etc.) that takes advantage of the speed of Fibre Channel and a standard server. With Avid Unity you can continue to add more drives to the network without disrupting the existing configuration. An administrator can change the size of any partition at any time without affecting the other projects on the drives. This means that you can get a much more efficient use from all of your storage. How many times have you had one room with extra drive space and another room that needs it right now? How many times has a project moved from one suite to another? Imagine if that changeover was as easy as assigning permissions to another editor for the same media.

The idea of sharing the media and metadata will eventually change postproduction more than any other single aspect of nonlinear technology. Having editors working in parallel with sound designers and graphic artists will improve efficiency, collaboration, and allow facilities to offer new services. That is just the beginning. Systems that can be configured as render stations, digitize stations, view and approve stations, and compression and DVD mastering stations are perhaps less than a year away. Any new facility should design all the workflow around central storage and begin to really get more value from owning multiple systems.

SNEAKER NET

Finally, the last, cheapest, and worst choice for networking: the Sneaker Net. If you have invested in the MediaDocks, this is not so bad. You just shuttle the media from room to room by hand (sneakers are optional). It means stopping work on two systems until the transfer is complete. If the goal is just to take everything to another location, it is simple and easy. Just don't trip! (Shoelaces should be tied!) If possible, don't fall into the common practice of disconnecting standard, fixed drives and moving them from room to room. You may get away with this for awhile, but eventually you end up with bent pins, kinked cables, and a few error messages. There are some extra complications with Windows NT 4.0 and striped drives, so carefully follow the Disk Mounter procedures outlined in Chapter 13. Make sure the person doing the Sneaker Net transfer knows everything there is to know about SCSI configurations; this may end up being you!

RELINKING

All master clips, subclips, sequences, and graphics must link to media in order to play. When you log a shot into a bin, you create a text file. When you digitize the shot, you create a media file on the media drive. The relationship between the master clip and the media file is considered a "link." If the master clip becomes "unlinked" from the media file, it is considered offline. The media may be on the media drive, but if the master clip is unlinked, it is considered offline and unavailable. In order to link, you must highlight the clip and choose Relink from the Clip pull-down menu. The ability to link, relink, and unlink is very important to sophisticated media management. We will revisit these concepts over and over again in this handbook. Here are ten rules that govern this complex and confusing behavior.

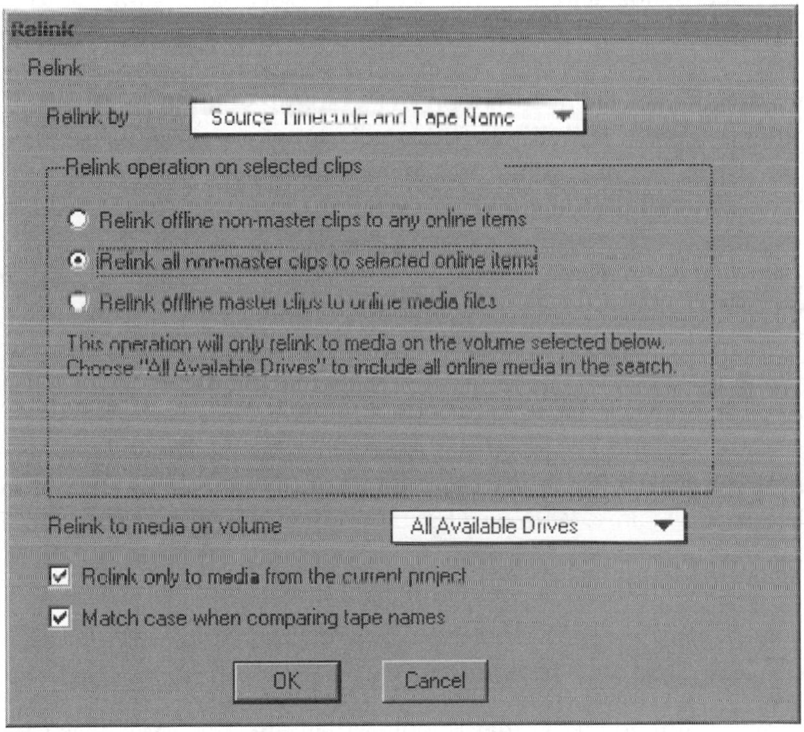

Figure 4.9 Relink to Selected

Rule 1 — Tape name and timecode

All linking between a master clip and a media file is based on identical tape name and timecode with media in the 5.x, 6.x, or OMFI MediaFiles folder. Symphony Universal has a relink by key number for picture only. This means you can relink media when you have a master clip, subclip, sequence, or graphic that is offline and you know that the media is on the drive in the MediaFiles folder. Highlight the object in the bin and choose Relink from the Clip menu. Your system searches the active MediaFiles folder for media that uses the same tape name and timecode.

Rule 2 — Tape name and project name

Just because the tape name looks the same doesn't mean it is the same tape. There can be only one tape logged per project with a particular name. Every time you add a tape as New in the Tape Name dialog, you are creating a unique tape that is associated with that project. You can use two different tape 001s in the same project, but they are logged in separate projects and are always considered different tapes.

In pre-7.x/2.x versions, if you uncheck "Only show this project's tapes" in the Tape Name dialog window, you see the project name attached to the tape name. Tape 001 becomes Project X:001. In recent versions, the project name is now displayed as a column next to the tape name in the Tape Name dialog window. You can choose to show other project's tapes if you think the tape you need is in another project. To reduce confusion, all tapes for a project should be logged in the same project or in another project with the exact same name.

Having said all that, there is a way to bypass project name using the new linking capabilities of the NT releases 3.0/9.0/2.0. In the Relink dialog, uncheck the "Relink only to media from the current project." Then the tape name only is important and not what project the tape came from.

Rule 3 — Size does matter

A master clip cannot link to digitized media that is more than a few frames different from the master clip's start and end times, even though it has common timecode and tape name. With 6.x/1.x, this rule was much stricter and the master clip needed to be exactly identical to the media file. For various reasons, this became looser in later versions, but only by a few frames.

Rule 4 — Subclips are less choosy

A subclip will relink to media that is longer than the subclip. This is true even if the subclip is exactly the same length as a master clip that will not relink. Subclips are programmed to link to more media than the subclip start and end times.

Rule 5 — Sequences are really subclips

A sequence may relink when the individual master clips that are contained in it will not. Think of a sequence as many subclips.

Rule 6 — Multi-part files make things more complicated

A sequence or a subclip will not relink to a media file that is shorter than the media it needs unless it is a multi-part file. In recent versions, master clips were able to have multiple media files because they were split up among several drives during the digitizing. This new feature allowed users to avoid the Macintosh two-gigabyte file size limit and use disk space more efficiently. It also made relinking a bit more complicated. If only one of several media files links to the sequence or subclip, then there will be black in place of the missing media.

Rule 7 — Relinking master clips is different than relinking sequences

The Relink Master Clip(s) checkbox in the older Relink dialog box must only be used with master clips. In the new Relink dialog, the system will gray out inappropriate choices. If a master clip and a sequence are both selected, then you must choose which one you really want to relink.

Rule 8 — Relinking a sequence to selected master clips only works in the same bin

In the Relink dialog box, the "Relink all non-master clips to selected online items" button can be used only when you have highlighted specific online master clips that you want to relink to a sequence. In older versions, this was called "Relink to Selected." There is no way to unlink a sequence, only forcing it to relink to other media or taking the original media offline.

This option works only within a single bin. It is primarily used to link a sequence to media at another resolution or from another project. To relink to clips from the Media Tool, the clips must first be dragged from the Media Tool into the same bin with the sequence. Everything in the bin to be relinked must be highlighted. Do not check this function unless you are specifically linking "to selected" or nothing will happen.

Rule 9 — Media databases affect linking

If all of your efforts fail to produce a relinking, make sure you are working with the version of the bin that was actually digitized (i.e., the same project). With the latest version you can uncheck "Relink only to media from the current project" and try again. If this still does not work, you will have to rebuild the Media Databases and desktop file. Your drives may be too full to correctly rebuild the Media Databases or you may have too many small objects and not enough RAM. You should delete unneeded precomputes or media to create space and force the Media Databases to rebuild.

Rule 10 — Media management demands updated Media Databases

When using any of the recent versions of the software, be very careful when copying files at the desktop level. Media databases are called msmOMFI.mdb and msmMac.pmr files on Mac, or msmFMID and msmMOB on NT. They must be the most recent files in the OMFI MediaFiles folder. If they are not, you may have to force them to update. Choose the Reload Media Database command under the File menu to force the system to look at all the media in the folder and to create a new database if necessary. Opening the Media Tool will do the same thing. If this still does not work, you may have corrupted databases and must throw them away. The system will automatically make new ones when you relaunch the editing software or click back onto an open window in the Avid software.

Although you needed to force the rebuilding of the media databases occasionally under the earlier releases, you will need to do it more often now — basically every time you copy something from one drive to another. This is because in a shared media environment (like MediaShare or Unity) your system would stop and rebuild the database every time someone else did something with media to any of the other systems sharing the drives. This would slow all editing to a halt.

If media is still offline after throwing away the Media Databases, then you must try the Relink command. Overall, you must be more responsible for choosing when to update your Media Databases yourself since, in the interest of speed, it will not happen automatically under all the same conditions. As long as you understand that the Media Databases must be updated (thrown away and re-created) or reloaded whenever you have a media management problem, you will be fine.

UNLINKING

Unlinking is one of those powerful, dangerous, useful, and poorly understood functions that people know they should use but don't really know when. Sometimes a link must be deliberately broken using Unlink. In the Clip menu, highlight the desired clips, hold down Shift-Control on Mac or Shift-Alt on NT, and Relink becomes Unlink. Any media that has been digitized for this master clip now becomes offline. The system considers this master clip has never been digitized and is not linked to any media *or to any sequences*. The sequence linking is important because, otherwise, every use of this clip in any sequence will be changed. You want to change this one master clip, but you probably don't want every use of it to change, too. That is why Unlink is required as a safeguard.

You can then modify duration of the clip, but you must redigitize all the unlinked clips. Do not unlink media that has no timecode and do not modify the master clip because you will be unable to batch redigitize.

Unlink is extremely useful for multicam projects. After batch digitizing Reel 1 from Camera 1, you can duplicate all the master clips, unlink them, and change the tape name (Reel 2, Reel 3, etc.). Now you can batch digitize all the other camera angles. Just be sure to duplicate the original clips using Control/Command-D and not Alt/Option-drag to another bin. You must be working with a true duplicate and not a clone of the master clip before you unlink.

There is no way to unlink a sequence using the Unlink command. Sequences have a loose link to media that allows them to change resolutions easily. The best way to unlink a sequence is to duplicate and decompose. You can throw away all the new decomposed master clips and just use the sequence. Because a sequence is loose about linking, you don't really need to unlink most of the time. You can just force the sequence to link to new material ("Relink all non-master clips to selected online items" with both media and sequence in the same bin), and it will automatically break the links to the old media.

NEW RELINKING FEATURES

Knowing full well that linking and relinking are rather complex, Avid has added some controls to make it less confusing and simultaneously more powerful. With the NT releases 3.0/9.0/2.0, there is a new Relinking dialog window. This window will allow relinking based purely on tape name and timecode, and will bypass different or missing project names. Because this may occasionally cause the system to link to the wrong tape, there have been some added safeguards. You can choose whether the media is from the same project and you can make sure that the tape name is spelled exactly the same. If these two new choices are activated, then the first time you try to relink nothing may happen, however, if you are relinking to a terabyte or two of media, you will appreciate these choices when working with the tape names of dozens of projects online.

USE COMMON SENSE

So while there is a lot to be in charge of in administering an Avid system, much of it is common sense and taking advantage of existing computer peripherals and software that make your job easier. Whatever policy you decide on, make sure it is followed uniformly. Make sure that all members of your staff are educated on the correct procedures as well as just a little troubleshooting. Then they can deal with those questions themselves during the night shift. If set up right, you will significantly reduce downtime and make it easier to diagnose and solve technical problems and missing media.

5

Long Format and Improving System Performance

Whenever you push a system to the limits of its capabilities, different rules of operation may apply. All of the functionality described in this book should operate correctly no matter how many drives or how long the sequence. But any time you work with extremely large files on a computer, you want to minimize the wait times involved with the extra work all the hardware must do to keep up. As with all programs that give you an advanced level of flexibility, the Avid allows you to work only on the specific parts of the project you want to change. If you are working on a sound mix, you may want to minimize the video or vice versa. This chapter describes various techniques developed specifically to increase performance and to cope with situations that occur only when working on longer format pieces.

Any sequence longer than a few minutes that requires more than just a few drives for the media could be considered long format. Only as the number of drives, master clips, and the length of the sequences creep up will you experience any benefit from many of these procedures. Many of these techniques will speed up the response of the systems no matter how long the project.

If you experience a sudden slowdown in performance, you may be experiencing a simple technical problem that can be troubleshot and solved quickly. Generally, a sudden slowdown occurs if lots of very large bins are being opened and closed, or other pieces of software have been launched and quit throughout the day. This is just normal RAM fragmentation and can be solved easily by shutting down the computer and restarting. Most people do this before the beginning of a shift, and some, working with lots of media and bins, also shut down at lunch and then restart. Dismounting drives that hold unneeded media can instantly reduce the amount of objects the system needs to account for. This can be done on the Macintosh by dragging the icons to the trash (or highlighting them and using Command-Y). You may want to rebuild the desktop and the Media Databases as well. On the NT you can dismount drives using the Disk Administrator. Details on those procedures are covered in Chapter 9.

The absolute first thing that must be considered before starting a long project is how much RAM is in the system. Over 150 megabytes of RAM is not too much. The more objects or digitized master clips, precomputes, sequences, subclips, etc., the more RAM you need to handle them. As I have mentioned before, any cost for RAM more than pays for itself as the project gets longer and more complex. The only bad side effect of having so much RAM is that the startup process of the Macintosh takes considerably longer; sometimes a few minutes before you even see the happy Mac icon. All of that RAM must be initialized, which takes time.

BUILDING THE PIPES

Every time you press Play, there is a preloading process where all media in the sequence must be scanned and analyzed. If there are mismatched resolutions in monitored tracks or mismatched sample rates in the monitored audio, you may get an error message and the sequence will not play. You may see the video slate "Wrong Format" when trying to play any resolution that could not mix with the resolution of the very first shot of the sequence.

The benefit of this preloading or "building the pipes" is the ability to play the entire sequence backward or forward at any time. The drawback is that the system is working very hard to provide access to all your media, although you may only want to see the next 30 seconds. When the sequence gets quite long, there may be a noticeable delay after pressing the Play button.

There are three methods to reduce this delay. The first method is to work with shorter sequences. This has many benefits, including not having to worry about sync 20 minutes down the Timeline when you just want to make a few quick changes to the opening montage. Focusing on a specific trouble section and massaging it without worrying about affecting the rest of the sequence cannot help but speed up the editing process, even disregarding technical playback limitations. Many projects lend themselves to being broken into smaller pieces anyway. Films break easily between scenes, episodics break between acts, and corporate programs and documentaries break at location or subject changes. When it becomes necessary to view the entire project at once, it is a simple process to Shift-click all the smaller sequences and drag them to the record monitor where they instantly become a composite master. Perhaps a fade to black or dissolve needs to be added between these segments, but otherwise you have a quick, rough assembly of the entire show.

After the approval viewing or the digital cut is over, go back to the individual segments to make the changes and archive the composite master. There is nothing worse than making changes on a short, individual segment but then showing the older, composite master sequence without those changes!

The second method is to only Play to Out (Alt-Play or Option-Play). The system builds the pipes only up to the marked-out point. This approach will be necessary at a later point in the project, after all the segments have been combined, or if there is no way to break the sequence into smaller pieces. If you get into the habit of quickly jumping down the Timeline a minute or so, marking an out, and then using Play to Out, you get a faster response from the system.

PLAYLENGTH

The third method to reduce playback delay is called Playlength. This is a button in the Command Palette (PL) and can be mapped to the record side of the source/record mode or mapped to the shifted function of the Play key. When you activate Playlength, the play button dims down the normal black icon to a gray icon. When active, Playlength restricts the building of the pipes to one minute in front of the blue position bar and one minute after the blue bar. When it reaches the end of the one minute, it stops playing and you need to press Play again. Because the preload is so much shorter, the response time of the system for much of its operation (including trims) is noticeably snappier.

Playlength does not affect the Digital Cut. This prevents a possible embarrassing situation when you go to tape, or as some very brave souls have tried, straight to air.

Figure 5.1 Specifying Playlength in the Console

If you do not have a model with the Playlength button, then you can access this feature through the Console. The Console is powerful but potentially dangerous because it is designed primarily for the programmers to debug the program or run various tests. Here is the obligatory warning. Any command typed into the Console is completely unsupported by official Avid policy. Under the wrong conditions, you can permanently change the way your software operates and you may have to reload it from the original disks to get it to behave correctly again. Don't mess around with the Console unless you know what you are doing!

Despite all these warnings, here is how you can activate Playlength without the button. This will change the length of pipes that are built using the Playlength button:

- Open the Console under the Tools menu.
- Type "Playlength 2." Notice the space before the number.
- Hit Return. On newer systems, you will get a message confirming what you have done.

Since the use of the Playlength button defaults to two minutes, you may want to change it to make it longer or shorter. When you launch the application the next time, it will always go back to the two minutes, but it will stay at the new length for the duration of the session.

To return the system back to normal without the Playlength button:

- Go back to the Console.
- Type "Playlength 0" and hit Return.

Quitting and relaunching the software will always return the system to normal Play behavior.

RENDER ON THE FLY

Other ways to improve performance are to utilize techniques to reduce the amount of media that must be retrieved, played, or rendered. There is a menu choice under the Special pull-down menu called Render on the Fly. Render on the Fly is a temporary render of one frame performed in the RAM when you place the blue position bar over the effect in the sequence. Nothing actually goes to a media drive. With Render on the Fly turned off, when the position bar stops over an unrendered effect, the system does not try to composite that single frame for viewing. The system shows the highest numbered track that does not contain an effect. This means you can scroll through the Timeline and stop in the middle of an unrendered effects sequence without the delay involved in previewing all the layers. You can move quickly to perform tasks that do not require seeing these composites, including audio mixing, trimming to marks based on audio, and moving segments around the Timeline.

Editors who turn off Render on the Fly on a regular basis end up mapping it to a function key. The only drawback to disabling Render on the Fly, besides not being able to see every layer, is that some objects, like titles or imported matte keys, are considered "effects with source." This means that they must be rendered on the fly to be seen. If Render on the Fly is disabled, you see only black when loading a title into the source window. Ironically, you can view them as soon as you hit Play because then they are rendered in real time. Leaving Render on the Fly disabled has resulted in a few calls to Avid Support, so remember to turn it back on before handing the system over to the next editor.

With the latest NT versions 3.0/9.0/2.0, Render on the Fly does not take effect on non-real-time effects until you release the mouse. Since non-real-time effects (blue dots) are much slower to render, this noticeably speeds up scrolling through the sequence. If you need to see these effects, you now have control of when they Render on the Fly by releasing the mouse.

There is also a checkbox for enabling Render on the Fly in the Trim Mode under the Trim Setting. It defaults to being off, but once turned on, it allows you to see multiple unrendered layers and effects while in the Trim Mode. This slows trimming considerably but may be necessary to make content-based trims with those layers in view. If you find yourself needing this ability on a regular basis, make two separate Trim Settings and switch between them in the Project window. This kind of change is made more efficient using Workspaces and "Activate settings linked by name."

MOVING THE VIDEO MONITOR

Another way to improve performance while working with effects is to move the video monitor icon to a lower track. This means that since you are not monitoring those tracks, the media does not need to be retrieved and played. Again, you can move quickly through a multi-layered sequence without waiting for unneeded images to be recalled.

With later versions of Media Composer and Symphony, a user setting in the Timeline Setting allows you to turn off Auto-Monitoring. Auto-Monitoring is when the video monitor icon follows whatever you are doing in the Timeline. If you patch to a higher channel, even if you have turned the monitor off, it will activate on that top track and force a longer Render on the Fly. With Auto-Monitor turned off, you are responsible for moving the monitor where you want it to go, but it will always stay where you leave it.

To move even faster, turn the video monitor off completely. Sound mixing and rearranging of large chunks of the sequence will speed up remarkably. This is also a good way to isolate whether playing back a video track is causing a particular error. If you stop monitoring the video track and the playback problem disappears, you have isolated the problem to that track.

RENDERING FOR SPEED

By occasionally rendering real-time effects during a break, you can make it easier for the system to cope with playing many short clips. The system is always trying to display all real-time effects by doing the compositing in the CPU. Playing back a precompute—a rendered effect file—takes less effort and can result in faster response time. If you are having problems playing a particular section of fast, short cuts, you can apply a submaster to the track above and render it. This puts all the short media files into one rendered precompute so the system needs to play back only one single file when you monitor the track with the submaster. When you make the EDL, you avoid the video track with the submaster and still generate a clean, accurate list.

VIDEO MIXDOWN FOR SPEED

If you don't need an EDL, you can create a video mixdown of the section. A video mixdown combines all the separate media files and precomputes and makes them one new media file and a new master clip. This also plays back very quickly with the added benefit that, if the segment has several layers, those media files required for the layers can be completely eliminated. If all the effects in a sequence have been rendered, then making a video mixdown is as fast as copying a file. This is an insignificant amount of time compared to the continued benefit of faster access time.

Even after you render an effect, the layers below the top track continue to cause media to be retrieved in case it is needed. With video mixdown, there is just one layer, one media file, and subsequently, faster retrieval time. Don't use a video mixdown if you need an EDL or you need to match frame back to an original source in the segment. A video mixdown has no timecode and retains no links with the original media; it is purely a convenience for simplifying a complex segment. Always make a duplicate of the sequence before cutting a video mixdown over the original layers so you'll have a non-mixdown version of the sequence to come back to for revisions.

MINIMIZE TIMELINE REDRAW TIME

There are ways to minimize the screen redraw time of the Timeline. This is especially important when many icons need to be constantly updated. You can go into the Timeline fast menu or Timeline User Settings and start disabling the display for various icons. By turning off effect icons and dissolve icons, you lose that visual information, but if you are positioned over the clip, you can still see the effect in the record window. Effect Contents can be disabled so that if there are nested effects, they show up only as the outside effect without trying to show all the layers inside. If you have not seen this setting at work, it is because it is visible only when you are zoomed in closely on the Timeline and you have enlarged the height of that track (Ctl-L or Command-L).

Never work with the audio waveforms displayed unless you specifically need them. This can be better utilized in later versions of Media Composer and Symphony where a Timeline User Setting shows waveforms only between mark inpoints and outpoints.

Any of the modes that require images to be drawn in the Timeline should be disabled, like Heads View or the Film track, and in general, any display you can work without temporarily should be turned off. I always leave sync breaks on, but I might turn them off for audio when I am editing primarily video. You should save a Timeline view with all these display features turned off so you can get to it quickly when you really want to move fast.

Figure 5.2 Four-Frame Display

An instance where redrawing images can slow performance occurs when using the segment editing arrows. As soon as you begin to drag a segment in the Timeline, the four screens showing all the transition frames being affected must be drawn on the screen. When moving especially fast, this redraw can delay the system enough so that it perceives a click and drag as a double-click on older CPUs. The system waits for you to release the mouse so it can display nesting, the command it has received from a double-click over an effect in the Segment Mode. It does not allow you to drag at all. In recent versions of Media Composer or Symphony, the four frames can be disabled as part of a User Setting under the new Timeline Setting (Show Four-Frame Display).

In earlier versions, this four-screen monitor can be turned off only for the session. Here is the old-fashioned way (version 6.5.x and earlier) to disable the four screens in segment mode:

- Click in the Timeline with a segment arrow.
- While still holding down the mouse, hold down the Shift button.
- Drag the segment back and forth slightly.

You hear an error beep and the four screens are turned off until you perform this procedure again or restart the system.

With the new Timeline User Setting in Media Composer and Symphony, there is also a command to turn off the segment arrow double-click that displays nesting. These Timeline Settings are welcome additions for the speed freaks out there who are always working a few steps faster than the system can display.

On Windows NT systems, another little trick gets the screen to update faster. If you go under the Start menu and choose Settings/Control Panels/Display you will see a series of tabs. Choose Plus! and uncheck "Show window contents while dragging." This will now give you just an outline of any window while you are holding the mouse down and moving it. Since the system doesn't have to redraw so much information, it can give you control and update the screen faster whenever you move windows.

Working with long projects means trying to squeeze all the extra speed out of your system. It means stripping down the features to the bare necessities to minimize screen redraw time and wasted time accessing images and effects you don't need at that moment. Many of these speed-based User Settings can be turned on or off using the new Workspace functions and mapped to a single key. In the end, you are working with the most flexible, customizable system for media management and maximizing the use of the drives and CPU. No other system comes close.

6

Importing and Exporting

A major benefit of working on a desktop system for video is the ability to import from and export to other third-party programs. Bringing in graphics from new and different sources is not only convenient and fast, but also it all but eliminates the traditional copy-stand work that was done in many online suites.

Importing and exporting is now much simpler. With graphics, the system takes an educated look and determines the format of the graphic. If recognized, the system automatically imports it. This removes a lot of the complications that resulted when you had to change the format of your graphic to make it compatible. This new ability is possible because Avid has incorporated the technology called HIIP (Host Independent Image Protocol). There is autodetection of file type using HIIP in current versions and there are now over 25 formats for import and export.

There are also the new drag-and-drop capabilities of the last few releases to help simplify and reduce errors. The user creates an export template based on the requirements outlined in this chapter and is assured every export will be consistently correct. By dragging a sequence or a master clip to the desktop level, even complicated exports can be done by beginners. The same HIIP technology allows users to drag a graphic straight from the network and drop it on the open bin. Preset, named, copied and carried, import and export templates will make interoperability with other software a one-step process.

Import and export procedures depend on the end result. Do you want the highest quality? You must follow some basic requirements for format, size, and resolution. Do you want to play the material on a CD-ROM? You must make some quality tradeoffs to get a decent throughput of the images and audio on consumer computers. You must, above all, educate your subcontractors and graphics department (even if that is just you!) and find a consistent procedure for acquiring the type of graphics and animation you demand for your clients and productions.

IMPORT AND EXPORT BASICS

Some basic issues should be understood when working with computers, graphics, and video. Because of the way computers have developed with their reliance on RGB color for their screens and memory, and the way video developed, beginning with black and white, the two mediums have never had a particularly easy coexistence. One of the hopes for high-definition television is that some of those issues will be resolved, but the addition of multiple, incompatible formats has rarely made things simpler.

COLOR SPACE CONVERSION

Computers and video work with different color formats and different kinds of scanning and scan rates. Computers traditionally work in RGB color space that is easier for displaying images on an RGB computer screen and working with computer memory. RGB has a more limited span of colors to choose from, but it is very dense, with many more possible points within that range of perceptible colors. The ironic part about camera-originated video is that it begins as RGB and is then converted by the camera to a different color space called YCbCr (or YUV). This new YCbCr color space spans beyond the range of visible colors but has fewer points within the visible range. Thus a conversion from YCbCr to RGB can force colors to change because a true exact match does not exist or because colors are out of the range in the new color space ("out of gamut").

The change of colors when the video camera's RGB values are captured and then converted to YCbCr as the signal is recorded to tape is a greater source of color change than anything that happens later inside the computer. The amount of color information is cut in half as you move from a 4:4:4 RGB signal within the camera to a 4:2:2 YCbCr signal on tape. So yes, there is potentially some change to your video colors during the digitize process on an Avid Broadcast Video Board (ABVB) as it converts the signal to RGB for editing. You will most likely never see it on a standard video monitor. The conversion never affects monochrome values no matter how many generations you go using the ABVB. The Meridien video subsystem works in YCbCr or YUV, so this is no longer an issue. If your colors or video levels are changing, it is more likely you are exporting or importing into the system using the RGB levels choice instead of CCIR levels. This chapter will discuss the details of these choices later.

The ABVB is the video board for the Macintosh Media Composer version 6.x and 7.x and for Xpress models up to 2.0 (also on the Mac). It is ITU-R.bt.601 compliant. (This regulation used to be called CCIR-601, but they changed the name to reflect another committee.) ITU-R.bt.601 compliant means that it follows the standards for digital video. This makes things a little more complicated because most computer graphics programs do not conform to this video standard. At the same time, it opens up some interesting possibilities.

The reason it is more complicated is that the early video boards (the NuVista+ included) and prosumer video boards, although some may be of good quality, do not capture the entire range of the possible video levels. A true ITU-R.bt.601 video signal can use a signal that is as black as possible without going into the sync section of the signal. It can also go above the brightest level that should be broadcast, giving some extra range for digitizing bright signals without chopping out detail. The other, older design boards use a more limited range, which in NTSC is from 7.5 IRE, which is the blackest you can legally broadcast in NTSC, to exactly 100 IRE. In PAL, the range can be shifted, but it usually goes from 0 volts to .92 volts.

Not allowing a video signal below 7.5 IRE means that you cannot use the levels all the way down to 0 IRE (superblack) for keying graphics in a luminance key. If the only way to use the graphic is from videotape and there is no mask, matte, or hi-con, you need to key the graphic with a luminance key. Digitizing or importing graphics and video above and below the broadcast legal limits means that you can bring an image into the system intact and then choose how you would like to change the levels later.

Another characteristic of the ITU-R.bt.601 requirements is that the frame aspect ratio is different from most graphics computer programs. With standard computer monitors, most people work with square pixel graphics with a frame size of 648 × 486 pixels in NTSC. This is fine if you are going to one of the older designed, non-ITU-R.bt.601 video boards, but the 601 standard size and the ABVB require the graphic to be 720 × 486, a non-square pixel size. If you are importing a graphic that is 648 × 486, this is not a terrible mistake, but it means that the import has to resize the graphic for you. This adds another step to the process and, if you have a lot of fine detail in the image, on pre-2.x/7.x Macintosh versions you may see some resizing artifacts. All of the current versions of Avid editing programs do an excellent job of resizing square to non-square, and many of these importing techniques become much simpler.

The final characteristic of graphics and video that must be taken into consideration is a video field. There are two fields to every video frame since a standard television signal scans the screen twice. Whether you are PAL or NTSC, you must use interlaced scan lines—every other line is part of a different pass drawing the image on the screen. Broadcasting this signal takes less bandwidth because only half of the image is transmitted at any one time, but it complicates things when you are taking an interlaced image to a non-interlaced medium like the computer. A computer uses a progressive scan, which means that one line is drawn on the screen, then the next, and then the next, on down to the bottom of the frame and then back to the top, and the process starts again. If you export an interlaced frame from the Avid editing application and there is some kind of horizontal motion in the frame, you see a difference between the first set of scan lines, field one, and the second set, field two. Although they are only a 60th or 50th of a second apart, you see jagged horizontal displacement of the image every other line.

All of these mismatches can be dealt with if you follow a set of guidelines that takes one kind of image and converts it to another. Sometimes there is a loss of quality because the transition is not perfect, but you can usually plan around these present shortcomings when using the mismatched formats.

Most images you have seen on television up until the last few years have, at some point, passed through a camera or been laid off to videotape before being edited into a program. With the ease of basic desktop editing and the combination of graphics that go straight from one computer graphic format to a computer video format, you have a set of new problems. These graphics have too much fine detail to reproduce well in the relatively low-resolution, interlaced world of video. Consequently, they buzz, flicker, and give us unpredictable results. A thin line may look fine on a progressive scan monitor, but the moment it moves to an interlaced scan medium like video, that line may be only one scan line wide. This means the line is only drawn on the screen every other field and causes a disturbing flicker. Images that originated with a video camera can never be recorded with that kind of problem. And since, in the past, most graphics were seen through video monitors as they were being created, they could be adjusted on the spot so that the design could take into consideration the limitations of what looked good on video. Colors were toned down and detail was blurred until the image was acceptable, and then it was put to tape.

Until all graphic workstations can figure out how to approximate what the final product will look like after being interlaced and reduced in resolution, you must be able to tweak the graphics after you receive them. The most common adjustments are to open the graphic in a graphics program on the editing workstation and add a little blur to areas that are buzzing with too much detail. If done with a little skill, the blur will never be noticed. In fact, a very slightly blurred image looks better than one that is too sharp. With Media Composer 7.0 and Symphony 1.0, painting tools allow you to do this without any third-party programs, but they must be rendered. Another common adjustment is to lower the saturation of a particular color or, in a worst case scenario, all the colors. Again, this can be done easily and safely using the Avid color effect or by running all the images in a batch through a program that allows you to create a desaturation macro like Equilibrium's DeBabelizer or Adobe Photoshop 4.0 or later.

IMPORTING REQUIREMENTS

There are three important factors to get right when importing images. They are size, resolution, and format. If you get these three characteristics correct, then any graphic should import correctly. With recent versions, most of these requirements become much simpler. In fact, there were such important changes in importing between versions on Macintosh that I am going to deal with the import requirements in three separate sections. The major changes were a result of changing from the NuVista+ video board to the ABVB and then changing to an HIIP-based import.

NuVista+ and Importing

The Truevision NuVista+ video board was used on Media Composer until version 6.0. All systems before that release need to get these particular statistics correct when importing graphics:

- Only PICT files will be recognized.
- All images should be 72 ppi (pixels per inch).
- All images should be square pixel sizes:
  - NTSC: 640 × 480
  - PAL: create at 768 × 576 then resize to 640 × 576 before import
- All video levels will be reduced to broadcast safe during import.

If any of these are not correct, then the image will appear distorted or will not import at all.

For ABVB systems before 2.0/7.0 on the Macintosh:

- Only PICT files will be recognized.
- All images should be 72 ppi (pixels per inch).
- All images should be square pixel sizes:
  - NTSC: created 720 × 540 then resize to 720 × 486 before import
  - PAL: 768 × 576 then resize to 720 × 576 before import and use Force to Fit
- All video levels will be mapped to ITU-R.bt.601 levels.

If any of these are not correct, then the image will appear distorted or will not import at all.

For all later editing systems on Macintosh or NT:

- Most common file formats recognized automatically (over 24 types).
- All images should be 72 ppi but would be imported correctly if not.
- Images will be resized to non-square during import and can be square or non-square sizes. Here are the correct frame sizes:

Resolution	Frame Size NTSC/PAL
AVR 12–77	720 × 486/ 720 × 576
AVR 2s–9s	720 × 243/ 720 × 288
AVR 2m–6m	352 × 243/ 352 × 288
20:1–1:1	720 × 486/ 720 × 576
15:1s–2:1s	352 × 243/ 352 × 288

- All video levels will be mapped to ITU-R.bt.601 levels.

All of the resizing is done automatically and at high quality. You can work completely in a square pixel format and the system will take care of the correct non-square size during import. File type is sensed automatically and imported correctly. As progress marches on, the process has gotten simpler and more fool-proof without reducing choice.

Aspect Ratio, Pixel Aspect: 601

Size and resolution are related. For the ABVB and Meridien video subsystem, the size must be 720 × 486 for NTSC and 720 × 576 for PAL. But if you are working in a square pixel graphics program like Adobe Photoshop, then your starting frame size should be 720 × 540 (slightly larger version of 640 × 480) NTSC or 768 × 576 PAL. It is only after all the creative work is done that the image is squeezed by resizing it to 720 × 486 (NTSC) or 720 × 576 (PAL).

When you import the graphic, the later versions will handle this unsqueezing automatically by choosing the Aspect Ratio, Pixel aspect: 601. Some versions have the choice as CCIR, but the new industry terminology emerged to refer to standard casually as 601.

With earlier versions of Media Composer, the graphic artist should do the squeeze first. They should squeeze the square pixel size to the proper non-square pixel size by hand using the graphics program. A squeezed image will appear to unsqueeze during import and will look normal for editing.

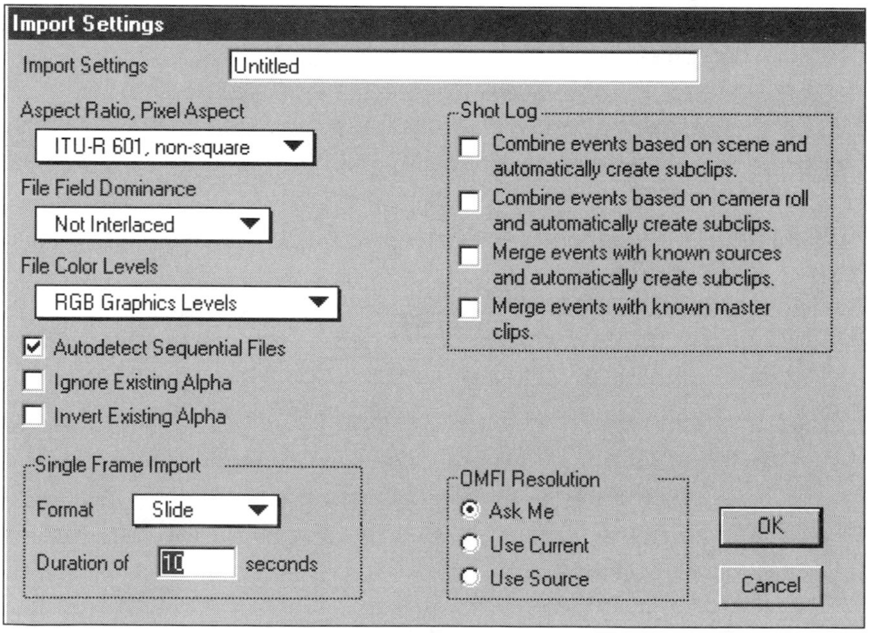

Figure 6.1 Import Settings

With PAL and the 1.x Avid Xpress and 6.5.x Media Composer, you must use Force to Fit as an import choice in the Import Settings for the unsqueeze to work right. The circles will be round and the squares will be square.

Different Frame Sizes

Under many circumstances you really need three choices for how to resize an imported frame. Also, the term Force to Fit implies that you are making something fit that was not done correctly. Properly sized images should not be forced to fit.

On older versions, the other two choices are Maintain, square; and Resize, square. These two settings work on square pixels and either keep the original aspect ratio (maintain) or take the image and enlarge it as much as possible without changing the aspect ratio (resize). You would use Maintain if you wanted to bring in a small logo or bug and stick it in the corner of the frame. If it has been sized correctly by the graphic artist, then you don't want to mess with it. Maintain the original square pixel aspect ratio and leave it alone. Resize will take that same graphic and expand it as large as possible. If the aspect ratio of the image is not correct for your video (4:3 or 16:9), then there will be black borders somewhere around the outside edge. Resize the square pixel image and leave the aspect ratio the same. Although these new terms are slightly more confusing than the old ones, they are more accurate and more predictable once you understand them.

The 1.x/6.x versions of the Avid Xpress and Media Composer allow you to Force to Fit, an option that takes any size and tries to stretch or squeeze to make it right. If you have the right aspect ratio it looks OK, but if it is long and thin, for instance, the only way to force it to fit is to squeeze it into the right aspect ratio. This is usually not acceptable and requires some human intervention to make the call on what must be cropped instead of stretched. The resizing of fine detail in pre-1.x/6.x versions will not look as good as if it was done by Photoshop or another third-party program that specializes in image resizing. The difference is subtle but worth it if there is a lot of complex detail and the finished quality is of the utmost importance.

If you have the wrong size and you do not choose Force to Fit, instead choosing Maintain Aspect Ratio, then the Avid system creates a black background around the image. The system makes sure that the largest dimension falls within the correct size. This still may not be acceptable if you have not taken into account the safe title and the safe picture area around the outside of the frame. Everything in the outside 10 percent of the frame will probably be cut off on most home TV sets. If that information is important, you must compensate for it before importing by pushing important details away from the edge of the graphic. You can also squeeze the image smaller after it is imported by using a picture-in-picture effect.

If you are working with the NuVista+ video board, things are simpler. You just import at 640 × 480 if you are working in NTSC and that is that. You don't have the wider range of video levels or the higher AVRs available to you, but it's hard to screw up. PAL unfortunately has another step with the NuVista+. Just as you must resize with the ABVB, you must resize with the NuVista+ with PAL. Start with the 768 × 576 and then resize to 640 × 576 before importing. Especially in PAL, you will notice a dramatic decrease in import time and increased image quality when you do the resizing outside of the Avid application.

So here again are the size rules for importing for 1.x/6.x:

- NTSC: With the ABVB, create the image at 720 × 540 and squeeze the image to 720 × 486 before importing. With the NuVista+, use 640 × 480.
- PAL: With the ABVB, create the image at 768 × 576 and then squeeze the image to 720 × 576 and import with Force to Fit. For the NuVista+, create the image at 768 × 576 and resize to 640 × 576.

16:9 Anamorphic Graphics

Many projects are now completely 16:9, so the graphics must be handled correctly to match the way the video is unsqueezed inside the Avid while editing. A 16:9 image is recorded with the video image anamorphically squeezed to fit the 4:3 video aspect ratio. A 16:9 monitor unsqueezes the image to make it fit the longer frame size.

If you are creating graphics or animations in NTSC:

- Create a composition with a size of 864 × 486, square pixel. Crop or resize to this size with the graphics program.
- When you output from the animation program, use the output module's Stretch function to resize the animation to 720 × 486. You can also nest the composition within a 720 × 486 composition. Use the Avid Codec.
- When you import your single graphic frame, choose pixel aspect ratio: ITUR 601.

In PAL:

- Create a composition with a size of 1024 × 576, square pixel. Crop or resize to this size with the graphics program.
- When you output from the animation program, use the output module's Stretch function to resize the animation to 720 × 576. You can also nest the composition within a 720 × 576 composition. Use the Avid Codec.
- When you import your single graphic frame choose pixel aspect ratio: ITUR 601.

Resolution

Resolution is linked closely to size in pre-2.x/7.x versions. In 2.x/7.x and later, pay no attention to the resolution because it will always be reduced to 72 ppi (pixels per inch). In earlier versions, if you have an image that is the correct size but the wrong resolution, it will import as if it is the wrong size. Graphic artists may insist that the higher the resolution, the better the image will appear in the final product, but this is not true for television above 72 ppi. It doesn't matter what the original resolution is, it must be converted to 72 ppi. If you have Maintain Aspect Ratio checked and the resolution is wrong, the older versions add a black border around the outside of the image.

Format

The last characteristic that must be taken into consideration is format. With older versions the only format that can be imported as a still and really work well is the PICT format. You can import directly from a Kodak Photo-CD and, if you have no other third-party programs, this is the only choice. But the size of a 35-millimeter scanned slide does not match the aspect ratio of video (square or non square) and so will give you a black border if you choose Maintain or Resize. It will squeeze the sides if you pick 601 or CCIR as it attempts to make the aspect ratio fit. You really must decide which part of this image to discard (not to mention if it is a vertical composition!); that should not be left up to the computer. Choose what must be discarded from the image and crop.

Alpha Channels

A good reason to work with PICT or TIFF images is because you can attach an alpha channel to it and use the higher quality anti-aliased edge of this alpha channel to create the best key the system is capable of. Other formats allow alpha channels, so there is less of a need to convert to PICT than in the past. But make sure, if you need an alpha channel, that the format you use supports it.

An alpha channel should be seriously considered as the method of choice for keying images. It is a fourth channel after the red, green, and blue channels of an RGB graphic. An alpha channel holds grayscale shapes that determine what part of a graphic will be opaque and what part transparent. It is a bit of work, carefully drawing around edges, and could easily be handed off to someone more skilled at drawing or who has a dedicated computer just for graphics. Once you become good at it, you may find it easier and faster to just whip one off yourself. Depending on your image—perhaps it is high contrast already—and your skill with the graphics program, you may be able to use some automatic selection functions (like eye droppers) to create your alpha channel.

Here's a quick tip for using the matte cutting tools in the Animatte effect in the Intraframe option in Media Composer and Symphony:

- Use Animatte to draw a shape in the Effects Mode.
- With the object selected, choose "Export Matte PICT" from the File menu and export to the desktop.
- Reimport the image to a bin and you have a real-time key.
- Step into the key once it is in the sequence and replace the static video with moving video.

You can also stay in the Animatte effect and be able to change that matte over time with keyframes.

Working with a graphics tablet makes it easier for the fine adjustments to a matte selection to be made in a graphics program. Make sure the edges you create are always anti-aliased since this gives you the better keying edge. Some people insist that anti-aliasing should never be added to an alpha channel because it gives a little unwanted border around the outside edge. This is a result of the three-pixel-wide edge that is created in a straight alpha channel when you anti-alias. It is part of why an alpha channel gives a nice, smooth edge that ramps from full opacity to full transparency. The three pixels cause the edge to extend one pixel beyond the edge that you have so carefully drawn. Luckily, it is very easy to adjust the edge of the selection and "choke" it back by the necessary one pixel. This is very easy in Photoshop 3.0 and later using Modify Selection, but it is tedious in Photoshop 2.5 and is well worth the upgrade. All

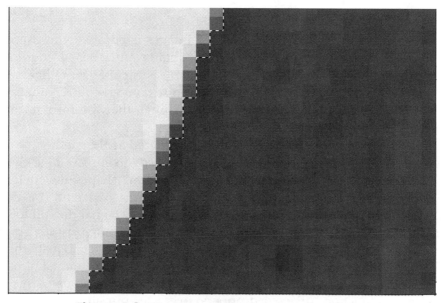

Figure 6.2 The Three Pixels of an Anti-aliased Edge

professional-level graphics programs should allow you to choke the selected edge by a pixel. This clears up any border that is a result of anti-aliased edges of a straight alpha channel.

The standard in television graphics on dedicated graphics workstations, like the Quantel Paintbox, has been to make any mask or hi-con area black where the keyed graphic was to be transparent and white where it was to be opaque. Most video switchers inverted this if necessary, but nonetheless this was common practice in the video world. In the film world, however, opticals were created with the opposite use of black and white, and this orientation, chosen many years before Media Composer became an online tool, now prevails. This means that, for simplicity's sake, all graphics should be created with an alpha channel that has white for transparent areas and black for opaque areas. Of course, if you know that the graphic was created in the older style, you can invert the alpha channel during import. Confusion arises when some graphics coming from one source are one way and graphics coming from another source are the opposite. This is something that should be specified as part of any facility's production guidelines. Do one test import to see if you have background and no foreground in your keyed graphic. If so, time to use the Invert Alpha Channel during import!

RGB VS. ITU-R 601 FOR IMPORT

We need to refer to the earlier discussion of ITU-R.bt.601 (CCIR) levels versus RGB levels when bringing in our graphic. Chances are, since the graphic originated in a graphics program, it was created with RGB values. This means that the levels of red, green, and blue at 0, 0, and 0 equal the blackest black possible in that program. Conversely, 255, 255, and 255 are the brightest. Now comes the dilemma: How dark and bright should they be in the video program? In the Avid software, you are dealing with the digital ITU-R.bt.601 values. During import, you are mapping the RGB values from the graphics program to another set of digital video values, the 601 specification. These values are the same for both PAL and NTSC.

The original graphic values, RGB, import the 0 levels from the graphics program to digital level 16, broadcast black. The 255 levels will be mapped to digital level 235, which is the brightest broadcastable white (both PAL and NTSC). This happens automatically during import when you choose RGB as your import choice.

The ITU-R.bt.601 values choice is sometimes called CCIR or just 601 in the Avid interface on different versions, so for simplicity many editors refer to the standard as just "601." With this choice, 0 from the graphics program is mapped to digital 0 and the 255 graphics levels are mapped to 255 digital video levels. There are legitimate reasons for both, which we will cover in the next few sections.

RGB Import

First, let's discuss the RGB values. RGB was the default, and the only way to import graphics up to version 6.5. It is still the only way to import if you are using the NuVista+ video board because it does not have the 601 abilities of the ABVB or Meridien. This is actually the simpler, safer way to import because nothing in the video signal will be below the legal levels for black. You are actually losing all of those levels of black if they were in the graphic, but you are gaining the confidence that the video levels will be safe for broadcast. It also takes all of the other values and maps them so that nothing is brighter than 100 IRE (or 1 volt in PAL). You cannot match an image from the graphics program at 255, 255, 255 (R, G, B) with an image that has a composite video level above 100 IRE or 1 volt.

ITUR.bt.601 Import

Importing using the 601 option takes the 0, 0, 0 RGB levels of the graphic and maps them to 0 digital. The darkest part of the picture is mapped to the darkest possible video level. The highest RGB value, 255, 255, 255, maps to 255 digital, the highest video level. This gets you into trouble during broadcast if you were not careful while creating the graphic in the RGB graphics program. The only way to avoid this problem is to run the graphic first through a third-party, safe-levels filter that makes sure that everything in the graphic is safe both in video level and color saturation. Sometimes these filters can change the hue as a by-product of this change. The Symphony SafeColor feature would fix this problem automatically if set correctly.

Sometimes you don't care if the video levels are safe, you just want them to match something already digitized. Clearly, importing at RGB levels changes your levels and importing the 601 levels does not, so after you have finished tweaking a shot you may want it to fit seamlessly into the existing video. The 601 option would be the way to go; 601 is of absolute importance if you are working with a frame of video exported that has values below or above the legal broadcast ranges. Some editors would say you should just redigitize the video again and get the levels within the proper specifications, but let's face it, that is not very often a real-world scenario. The amount you are over or under RGB levels may be tiny, but still visually noticeable if the match frame from the graphic to the video is not perfect.

Another reason you might not care if your video levels are safe is if you want to use the superblack for luminance keys. Using superblack allows the graphic or logo to contain black in the design and still be keyed using a luminance key because the level of black that is used to "key out" is at 0 IRE. This allows any 7.5 IRE normal black levels to "key in" or remain visible. To get a good key, make sure that nothing in the image you want to key is even close to the 0 black of the background that will be keyed out. I have seen people using between 11 IRE and 15 IRE in NTSC on graphics that have subtle shading. So, if you want something to be imported exactly the way it was exported or you want to make use of parts of the video signal beyond normal broadcast ranges, use the 601 values choice during import.

EXPORTING

There is certainly a joy to working on a general purpose computer instead of a dedicated graphics workstation. You always have the ability to quickly export a frame from the video application, tweak it in a graphics program, and import it back. No longer do you need to call the graphics department at the last minute to make the minor changes that are inevitable as the deadlines get closer. It makes it easy to take any frame and use it as a background, do some simple rotoscoping (painting on the video frame by frame), or isolate parts of a frame with an alpha channel. The trap, of course, is that you will always be counted on to do this once you have shown how easy it is!

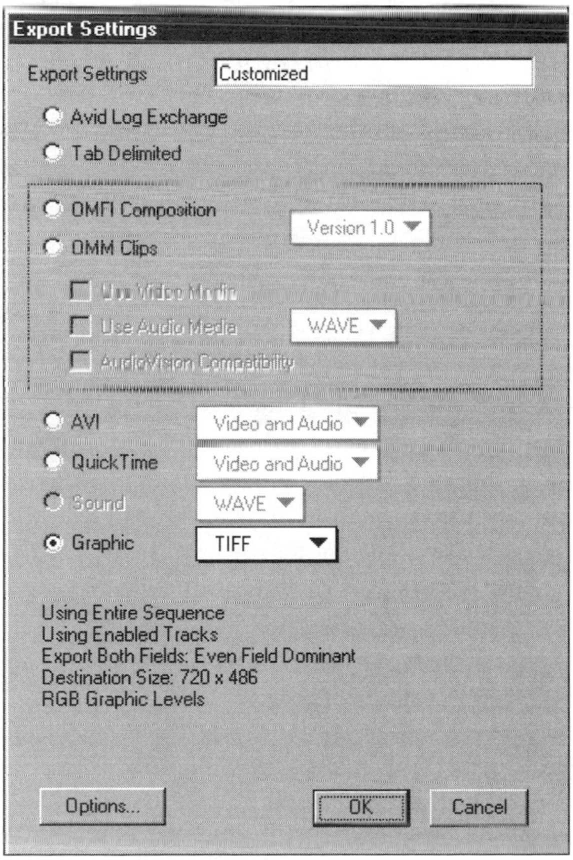

Figure 6.3 Export Settings

Export Templates

Since most exporting is done with only a handful of the potential formats, most people find the export types that suit them and ignore the rest. There is also a pattern to the type of exports. It makes sense then to take the settings that are used in exporting most often and save them as templates. You can create and save these templates as User Settings so you can take them with you from job to job with the assurance that you will always get the export right if you make the templates right. It is also great for experienced editors to make them for their less experienced colleagues or assistants to be used with an extra level of confidence that they will be correct.

Export Basics

Exporting has its own set of choices, but most of them are decided by the ultimate use of the image. If you are exporting an image to work on and then you reimport back, you want the export and import to be exactly the same. This means choosing 601 video levels and the native frame size. In NTSC that native frame size is 720 × 486 and in PAL 720 × 576.

If you change the frame size during export, two things happen. First, it takes much longer to export since the application must do more work resizing. Second, and more seriously, the scan lines are disturbed from their precise, standardized relationship. If you export a still at a small size for use in a document or for a Web page, the scan lines are not as important because the image is not going back to full frame video. A video import back to the Avid software needs the interlaced scan line information to reproduce the image exactly. If you resize, even to a square pixel size like 720 × 540 or 768 × 576, then when you bring the image back to a non-square video board, the import process does not know exactly how the original scan lines were laid out.

During import, the software puts the lines in a slightly different order from the original if the size has been changed. Forget trying to match back seamlessly to the original video if the scan lines are in slightly different places. You will also experience a loss of resolution because some scan lines are doubled to make up for missing ones. If you are exporting to reimport, always choose the native frame size and do not resize the image at all in the graphics program.

DE-INTERLACING

As I mentioned earlier, there is a slight difference between the image in the first scan line and the second scan line a 60th or 50th of a second later if there is motion in the frame. This is usually unacceptable in an exported frame if the end result is a print or designated for Web use—some kind of a still. One field must be eliminated, leaving only half of the resolution. Since every other scan line must be doubled or interpolated to fill in the empty space left by deleting half the frame, the image appears to be somewhat softer. Photoshop has a de-interlace filter that nicely removes your problem. Photoshop version 4.0 and later allows you to program this de-interlace as an Action or macro. For Web use, you are going to be reducing the quality of this image anyway in order for it to load across the Internet with better than glacial speeds, but enough third-party programs out there blend scan lines, reduce video noise, and limit the color palette so that you should always plan on running your graphics through an extra, final step before their end use.

Another reason to deal with interlacing is when you are working on compositing or rotoscoping software. If you take the interlace video exported from the Avid and de-interlace in the third-party software, you will reduce image quality problems. Then you re-interlace before rendering and exporting back to Avid.

Using the new 24P standard on the Universal Editing and Mastering options means that you no longer have to worry about interlacing. Since you are working with a true progressive frame, you can work only on full frames. In NTSC you will save 20 percent disk space, skip all 2:3 pull-down frames, and always export sharper stills.

SAFE COLORS

There is a danger that anything you do to the image in the third-party program may extend the video and chroma levels beyond the legal broadcast range. The first answer is to use only colors and brightness levels that are already in the image through the eyedropper or any other gadget you have in the graphics program for sampling part of the image. Remember that digital-safe levels are 16 for black and 235 for white. There is usually a palette in the graphics program that you can open to allow you to read the RGB levels of any individual pixel if you are suspicious.

This technique is not going to help you if you have applied a special effect filter to your image because the filter is not really restricted by safe colors or brightness levels for video. In fact, most people believe the wilder looking the image, the better! This will be a good way to identify the look of the late 1990s, but for now it poses the problem of checking each and every image before importing back into the Avid. Both Adobe Photoshop and Equilibrium's DeBabelizer have safe-level filters. Many safe filters just lower the brightness level of the image, but since DeBabelizer works with both chroma or brightness levels, it is the better choice. Again, importing with Symphony's SafeColor enabled for imported graphics will fix any over or under levels during the import stage.

All of this works pretty well if you stay consistent and keep your expectations realistic. You will never get a great still for print from digitized video that even comes close to a real photograph. Never plan to get your print advertising from freeze frames exported from the Avid unless you plan to run serious filters over them for a look that has nothing to do with reproduction accuracy.

IMPORTING AND EXPORTING MOTION VIDEO

The next step for import and export is using moving video. Usually, people want to import animations created by a 3-D program or an effect sequence rendered from a compositing program. There is also the demand to export for Web pages, CD-ROM, or for material to be used in a compositing program like After Effects. All of the already mentioned procedures apply, including resizing and safe colors, but with slight differences. Also you must decide between two choices for formats: QuickTime or sequences of stills.

Let's look at importing video first. The choice of whether to render as Quick-Time, AVI, or PICT/TIFF sequence depends on several factors. The first and most important is: What format does your third-party application have as an export choice? Usually less complex programs on the Macintosh have only one choice — QuickTime. This does indeed make things simpler.

QuickTime

QuickTime has earned a mixed reputation because most people experience it as a postage stamp, pixelated image that drops frames to stay in sync with the audio. This is how QuickTime plays back on most Macintosh computers without help from SCSI accelerators and special video boards. Most professionals shudder when contemplating what their flashy, quick-cut, and heavy-effects sequence would look like when reduced to such pathetic playback performance. But you are not going to playback QuickTime! Not yet anyway. All you want to do with QuickTime is use it as a transfer medium from one program to another. Since it allows conversion without any further quality loss, it has become the de facto standard for transferring video in the Macintosh world. Try to play this and you get giant frames that update about every five seconds, but copy it to a drive or send it down the network and it imports clean and shiny, with no loss of quality, into your editing software.

A QuickTime movie is a single large file that must stay under the Macintosh operating system file size limit of two gigabytes. Two gigabytes seems pretty large, but if you are working with high-quality video — this means between 250 and 675 kilobytes per frame — you will reach that size between around 4.5 minutes and 3.7 minutes (approximately 1.65 minutes with uncompressed NTSC). Usually when a project requires a long QuickTime movie for playback, it is a much reduced frame size and resolution. This chapter will discuss that issue later; for now we are concerned with using QuickTime as a high-resolution transfer medium.

The Avid Codec

The key to getting a QuickTime movie rendered at high quality that imports quickly into the Avid is to use the Avid Codec. A Codec is a system extension that any third-party program can access because it is in a central location on the host computer. On the Macintosh, it is in the System folder along with other extensions. On the NT, it is automatically installed in the C:\WinNT\system32 folder and on Windows 98 should be installed in the C:\Windows\System folder. The Avid Codec is a ".qtx" file. It allows other programs to compress and decompress a rendered QuickTime movie using the Avid media file format. This means that even though you are rendering a QuickTime movie, you are really creating an Avid media file (OMFI media in later versions) that is inside the QuickTime format. Technically, this is called *encapsulating*. To work best, use the native frame size and do not mix resolutions. We will discuss OMFI at the end of this chapter.

So make sure that you load the Avid Codec into any Macintosh or NT system you are using for graphics or animation and you will be given Avid resolutions when you choose the quality of your rendered QuickTime movie. Send the Codec to your graphics people, or anyone who is subcontracting graphics for you. And by all means, make sure it is the most recent version! You can get that information by calling Avid Support or by downloading from the Avid website (http://www.avid.com), and be aware that the version of the Codec may change on a different schedule from the software itself. The reward of using the Codec and native frame size is the almost real time import speed as you come back to the Avid.

Compatibility and the Codec.   Unfortunately, we must spend some time discussing what versions of Avid editing software use the QuickTime Codec and how because up until QuickTime version 3.0, QuickTime was not used on computers with Windows NT. Most people used AVI, which made conversion from one format to the other tedious and confusing. Combine this with AVI's lack of support for alpha channels, and you can see that there was much less interoperability in the bad old days.

First releases of Avid's NT editing systems did not support QuickTime since they were developed before QuickTime 3.0 was available. Instead, there was support for an AVI Codec that could only be used on other NT systems. By the time Symphony 1.0 and Xpress 2.0 shipped, it was a different story and support of some kind was necessary. For these first versions, a QuickTime-to-OMFI converter was included on a Macintosh-formatted floppy. Macintosh users would convert the QuickTime to OMFI (which is platform independent) and then import the OMFI media files. Although this was quick, it was definitely an extra step and as a workflow it was, as the engineers like to say, "sub-optimal."

Then an NT Codec was released, but it was not of use on NT Avid editing systems. If you are confused now, you are not alone, but it gets better — really. This first NT QuickTime Codec was for those who were creating QuickTime movies with NT versions of Adobe After Effects or other third-party compositing or animation programs. Render and create a QuickTime movie using the Avid NT Codec and then send it over to the Avid editing system on Macintosh. For those who were editing on NT, AVI was still the way to go.

Finally, in September of 1999, Avid released the next version of NT editing software (versions 3.0/9.0/2.0), which could now import and export QuickTime 4.0 using the latest Codec. There is a Codec for Macintosh and NT, and QuickTime movies are interchangeable. Life is good. For those still confused, you can download the latest versions of the Codecs from the Avid website (http://www.avid.com) and distribute freely to those in need.

So here's the list:

- Macintosh editing systems use QuickTime Codec 8.0.2 or later. This Codec supports alpha channels but the QuickTime-to-OMFI conversion will strip it out. All NT systems that support QuickTime also support alpha channels, so you can take the cross-platform QuickTime movie directly to NT and it will open just fine with the alpha channel intact.
- Macintosh third-party programs for NT Avid systems should use Codec 9.2 or later. These will support an alpha channel if your NT Avid system supports it (3.0/9.0/2.0).
- NT third-party programs rendering for use on Macintosh use NT QuickTime Codec 8.0.2 or later. If the movie has an alpha channel, versions of Media Composer or Avid Xpress that do not support alpha channels will import slowly.
- NT systems Symphony 1.1, Xpress 2.1, and Media Composer 8.0 should use the QuickTime-to-OMFI converter on a Macintosh or use AVI.
- NT systems Symphony 2.0, Xpress 3.0, and Media Composer 9.0 should use Avid Codec 9.0 or later.
- Progressive resolutions are only supported in Codec 9.2. This Codec is for all NT versions that support 24P and is only supported for Mac Media Composer 8.1 and Avid Xpress 3.1.

Size

Size is still a consideration for importing QuickTime and for all the same reasons. Before, with a still image, size was important because of the scan-line relationship, which is still crucial. But there is now another consideration — speed. If you render at anything other than the native frame size, you have to double lines to make up for the difference or throw away resolution and waste rendering time. You are also wasting import time since the Avid must resize each frame on the fly, which adds a considerable, unacceptable amount of extra time to the process.

If you must eventually create a square pixel frame, consider resizing using another program like Media Cleaner Pro on another computer (with the Avid Codec loaded). At least you are not tying up the editing program while the QuickTime movie is resizing.

The other choices involved, when rendering to import to the Avid, have to do with field order and field rendering. To get the absolute best-quality rendered movie, choose field rendering if your application makes it available. Field rendering takes longer to render, but it is worth it if you are working with complicated video and lots of detail and you want this to be a finished product.

Importing with an Alpha Channel

With current versions of Avid editing software, you are able to import QuickTime movies with an alpha channel attached (Codec version 8.0 and later on Mac, and 9.0 and later on NT). This is sometimes referred to as the 32-bit Codec because an RGB file with an alpha channel takes up 32 bits of information (24 for RGB plus 8 for alpha).

With the addition of a 32-bit QuickTime Codec, Avid has added a new choice for rendering and importing QuickTime movies with third-party compositing applications. In the past, there was much value in setting up composites to render in two separate passes on the compositing software. The first pass would be to render the foreground, and the second was to render the moving matte. Then the user would import them as two separate fast imports and layer them on top of each other for a non-real-time matte key; however, QuickTime supports the inclusion of the moving matte and the foreground as a single movie. Since 24 bits are used to represent the color and detail of the foreground, the extra 8 bits are used for the 256 levels of luminance for the moving matte or alpha channel. So a 32-bit QuickTime Codec allows you to render the alpha channel and the fill or foreground as a single movie and import it as a fast import. This makes QuickTime movies easier to move around (although bigger files) and simpler to use since the effect comes in as a premade non-real-time matte effect.

Older Methods with QuickTime Alpha Channels. On older versions (pre-2.5/8.0/2.0), importing QuickTime and using an alpha channel for compositing was a more complicated process. This section deals with the older methods of importing QuickTime with alpha channels. You have two choices if you want to import an animation with a moving alpha channel on the older systems.

The first choice is to not use the Avid Codec, but instead choose two other compression formats that support alpha channels: Animation and None. Even though you can load the new 32-bit Codec on the older systems, these versions of the software have no clue what to do with the alpha channel. They should import, but very slowly, and you will lose the main advantage of using the Avid

Codec in the first place. When you choose which type of compression to render your animation or composition, there are many confusing choices, but you should use Animation or None for finished quality work if you are not going to use the Avid Codec. Animation is designed for what it is named, animation. It uses a lossless compression if you adjust the quality slider to its highest level. Because it is compressed, it saves a little bit of disk space, and you will not hit the two-gigabyte file size limit quite so quickly.

The other choice, None, means that there is no compression at all. This results in larger file sizes and longer transfer or copy times, but there is nothing between you and your video.

When you import these QuickTime movies with alpha channels into the Avid, you end up with a two-layer matte key effect that must be rendered. This requires a relatively long render time once imported because the matte key is not a real-time effect and cannot take advantage of the fast render capabilities.

The other choice for QuickTime animation with an alpha channel is to use the Avid Codec but render the animation as two separate files in the compositing program. Render one pass for the animation and another for the alpha channel. If you can set them up to render one after the other and you are going home for the night anyway, the extra rendering time may make no difference to your schedule; however, you will still pay a time penalty when it comes time to render after importing into the Avid. You must take these two separate imported animations and layer them on top of each other, putting the alpha channel on top, and then apply a matte key effect to the alpha channel. It takes only a few seconds to assemble this effect, but the rendering time for a matte key is considerable.

AVI

Just a quick word about AVI as it becomes less important. This Windows-based video format filled the void in the Windows world until QuickTime 3.0. QuickTime 4.0 with the Sorenson Codec and other video streaming benefits has essentially eliminated it as a top choice for video interchange. AVI also lacks support for an embedded alpha channel. This means you must still render two versions of your animation—one for foreground and the other for alpha channel; however, it is still the only format for import or export on the first NT Avid editing systems (2.1/8.0/1.1), so occasionally you will need to convert from one format to another. My advice is to upgrade to the next Avid releases if you do a lot of third-party interchange and use QuickTime.

PICT or TIFF Sequences

Most people choose QuickTime as their medium of choice for animation because it is just one large file that is easy to keep track of and because it is the early universal choice of most animation programs. There are times when the animation program may originate on a computer platform that doesn't support QuickTime, such as a Sun Workstation, or you are choosing to work in a graphics program to affect each individual, exported video frame. In these cases you want to work with a PICT or TIFF sequence. If you are using Avid's Illusion, however, you can output as OMFI and simplify the process.

There is a little twist involved in getting these stills sequences to import to the Avid, which has to do with file naming. Any PICT or TIFF sequence must be recognized as a sequence and not as a series of unrelated stills. The way to be sure that you have a PICT or TIFF sequence is with the file name, which must follow a strict format of: Name.001, Name.002, Name.003, etc. The name must be followed by a period and a number. Before versions 2.0/7.0/1.0, "Name.001.pic" was an invalid name and the "pic" needed to be removed. With current versions, the extra part of the name after the number is accepted correctly. It must still be removed before import for earlier versions. With the earlier versions, you must get the PICT file name right or you will be faced with running every file through a renaming program, like Knoll Renamer, or creating a macro in DeBabelizer. These tasks are not extremely time consuming, just tedious and distracting from the work at hand (especially if you must track down one of these renaming programs while the clock is ticking).

The system will look for ".001" to be the first image in the animation. If you have started with ".000," then that frame will be skipped when the sequence is created. If there is no ".001," the system does not see the rest of the images as a sequence. With current versions, the system will autodetect sequential files if a series of files with the same name begins with ".001." Occasionally an individual file that is not part of a sequence is named with a ".001" and will be misinterpreted and imported as a one-frame file. If you do not want the system to assume this image is a very short sequence, you can turn this option off in the Import Settings (or rename the file). The finished PICT or TIFF sequence should come in as one video frame per image.

The main benefit of working with PICT or TIFF files to import into the Avid is that these file formats support an alpha channel. Importing automatically sets up a matte key effect, although, once again, because it will be moving, the matte key effect is not real time and must be rendered. This method is simple and relatively problem free unless the animation takes up thousands of stills. The Macintosh operating system may slow down to a point that it will be impossible to open a folder with too many images inside. You may want to break any animation into smaller sequences if you are concerned about filling a folder on the desktop with thousands of files.

Remember that the alpha must be inverted from what most programs create as a default. Remind your animator or invert on import. Either way, make sure you know whether you have white on black (need to invert) or black on white (import without adjustment) before you import so you don't do it twice!

Consider how easy it is to set up a network between an SGI and your Avid as an Ethernet or faster TCP/IP. TCP/IP is a protocol that you can use between two computers in your facility (technically, an intranet if done this way). This simplifies the transfer between computer platforms admirably and you no longer need to worry about a removable drive, compatible hard drive formats, or the time to copy and restore. Besides, most SGI workstations don't use a floppy drive.

The original source for your PICT or TIFF sequence may have come from the Avid. In this case, you are dealing with many individual PICTs and may be using these to go to an SGI or Windows NT-based program that doesn't use QuickTime (QuickTime 3.0 was the first cross-platform version). Or you may want to paint or draw on each individual frame and paint out wires, boom microphones, or ex-spouses. This is time-consuming because of the huge amount of work involved in 30 or 25 hand-painted frames per second. It is the ultimate "fix it in post" and last resort for fixing problems created in the field, but the results can be magical if done right. This is exactly the kind of work the intraframe painting tools were designed for.

A simpler way to affect each individual frame is to apply a filter to each one in a macro-based program like DeBabelizer. Set up the macro to open each frame, run an effect filter, run a safe colors filter, and then save the new frame with a different name. Of course, After Effects can do this to a QuickTime movie, too, but there will be rendering time.

Now that the new Avid format, called AVX (Avid Visual eXchange), is available, this will simplify the filter process even more. You will be able to buy a version of your favorite filter that has been converted to AVX and apply it in the Avid like any other effect. Early Photoshop plug-ins for version 2.0 actually show up in the Effects Palette on some versions when you drag them into the Third-Party Plug-Ins folder (inside the Avid Supporting Files folder). Photoshop plug-ins took so long to render, even a single frame, that they were generally impractical. The faster processors of today's computers and the presence of rendering accelerators like ICE make the use of plug-ins within the Avid much more viable.

When creating graphics from scratch to be imported into the Avid, remember that they must be resized to the non-square pixel format if you are working in a square pixel graphics program. Ideally, the animation program you are using works with non-square pixels, but if not, you must build every frame in the square pixel size and resize before saving. If you have been handed a square pixel animation by a subcontractor, you should resize it yourself. Again, a macro-based program like DeBabelizer or Media Cleaner Pro will come to your rescue. Alternatively, you can place the square pixel animation in a D1 size composition in After Effects, resize there, and render (don't underestimate the rendering time).

Media Cleaner Pro and CD-ROMs

The Avid allows you much flexibility with exporting QuickTime, but depending on your needs, you may still want to use a third-party program like Media Cleaner Pro by Terran Interactive to do the final compression for playback. Cinepak has been the most common form of compression for multimedia, but it is fast being eclipsed by the Sorenson Codec, which is now available using QuickTime 3.0. Sorenson allows a better image quality with a much lower data rate, especially for faster processors. It is more popular for low bandwidth uses like the Web. From the Avid, you can export at different frame rates, frame size, and compression choices. You could play back QuickTime straight from the Macintosh desktop or go straight to CD-ROM. Except for short movies, you may want to export everything using the Avid Codec at full frame rate and native frame size. Send the QuickTime file to another Mac workstation where Media Cleaner Pro can churn away in batches and create the best Cinepak or Sorenson compressed version with blended frames and compressed audio.

The exact procedures for creating the best CD-ROM are too detailed to go into here, but if your job is to get the best CD-ROM playback from your system, it is well worth researching. One thing to keep in mind is the extremely limited ability of a CD-ROM to play back QuickTime—it is a very slow drive. You have a maximum rate of between 175 and 200 kilobytes per second for a 2X CD-ROM player. Yes, there are 24X players available (which will be outdated by the publication of this book), but you cannot count on these for wide distribution of a commercial CD-ROM. If you are distributing to a controlled, in-house group, then you may go for a higher playback rate. Some people even have the luxury of loading their video or multimedia presentations onto a central server and are limited only by the speed of the hard drive and the network itself. Media Cleaner Pro allows you to make a series of intelligent decisions and weigh the pros and cons of your quality choices and the playback performance.

Cinepak

Compression type is the most important decision besides the frame rate for determining the data rate of a QuickTime movie. Cinepak is useful because it not only compresses each frame like motion JPEG, called *spatial compression,* but it compresses over time as well, called *temporal compression.* Temporal compression requires reference frames, keyframes that are not compressed over time. The more keyframes you have, the better your video will look. The compression algorithm has determined that certain parts of the frame do not need to be updated until the next keyframe. If there are sudden, unexpected changes like a series of quick cuts or abrupt action, Cinepak creates natural keyframes when the majority of pixels in an image change. So why wouldn't you have huge numbers of keyframes, for instance, every other frame? Because that defeats the purpose of compression over time. The keyframes are a reference, so they take up more space and require a higher throughput. It is because there are all of these smaller, compressed frames that you save space with Cinepak.

Audio for QuickTime

Audio is also a very important consideration for finished QuickTime playback. If you are exporting to a compositing program like After Effects, you are better off using the audio just as a reference so that you know you are hitting the right moments while creating your effects. The problem again is with data rate. What will most users respond to: jerky and blurry video or compressed mono audio? The answer most of the time will be that the quality of the video is more important than the quality of the audio. Unless the sound is distorted or unintelligible, people are happy playing lower quality audio through the little speakers in their computer. Media Cleaner Pro offers a good compromise for compression called IMA (Interactive Multimedia Association) that can be played back from all but the oldest Macs.

A few issues have to do with volume level with exporting audio from the Avid. Again, evaluate whether your audio is important enough to double its data rate when the majority of listeners will be listening to it in mono anyway. Here is the best method for mono audio when exporting as QuickTime:

- Make a duplicate of your sequence.
- Pan the audio either all to the left channel or all to the right channel.
- Raise the audio levels as high as possible without distorting them.
- Make an audio mixdown.
- Delete the original audio tracks, leave the mixdown track, and export.

OMFI AND AAF

A cross-platform choice for importing and exporting that is now used extensively is OMFI (Open Media Framework Interchange). It is a non-proprietary format that Avid developed, and over 400 manufacturers and software companies are using it. The beauty of OMFI is that it can contain important information about your Avid sequence. It can contain both media and composition information and can represent a much greater complexity than an EDL. It is the way that other systems input all of that complicated layering, effects, audio mixing, and links to digitized media. It keeps you from having to reproduce it all. If you are going to be exporting sequence information with your media, this is the way to go. Any facility that is looking to exchange large amounts of media across platforms and across applications should seriously look into using OMFI. The standard edit decision list is almost useless for representing complicated effects information across different manufacturers of hardware and software.

The main confusion about OMFI is that people forget that it can hold both composition information and digitized media. If you export a composition, then another system can open up all your layers and effects and batch digitize the media at the new resolution. If you export the media, then you must be sure that the other system can play that media file type. There have been several media file changes over the years with Avid systems. Double check before you export media.

Lately, a series of third-party programs can read and write OMFI, which is a very good thing. This means you can skip any intermediate format, like QuickTime, when creating animations for the Avid systems. Programs like Eyeon's Digital Fusion and Puffin's Commotion read and write OMFI media today. At Siggraph 1999, ICE announced that it was accelerating Adobe After Effects, and even before that, ICE programs like Commotion and ICE Blast used OMFI as a result of their ICE acceleration. This trend toward making OMFI just another media format will simplify interoperability as much as if you were using QuickTime native applications. Don't forget that you probably still want to work on copies of the original OMFI media files so any benefit of direct import and export will be tempered by the time to copy the files.

AAF (Advanced Authoring Format) is the new emerging standard for interchange. It will be backward compatible with OMFI and will continue to better represent effects and complicated sequences for multiple equipment manufacturers. You should begin to see AAF-compatible software beginning in the year 2000.

Using OMFI for Pro Tools

OMFI is used most often for moving audio from a video session on the Avid to a digital audio workstation. If you are moving the audio to a Digidesign Pro Tools session, there are two methods to consider. The first method requires less drive space and is faster, but the second method allows more flexibility. Consider the second method if you are going to use software that does not recognize the Sound Designer II format that Pro Tools and Avid Macintosh editing systems use. AudioVision supports a direct translation from the Pro Tools session, so once you convert it for Pro Tools, AudioVision can read it, too. Recent versions allow an OMFI export as an AudioVision-compatible file to skip this middle step.

The first method is to hand over the drive with the audio media files on it and create an OMFI file that is composition only and an SD II format. The OMFI holds all the edit information, and the drive containing all the audio files must be moved from the ATTO SCSI chain to the internal SCSI chain. You may want to consolidate your sequence before you do this if you want to put all the audio media on another drive for transport to the audio workstation. This allows you to keep working with the audio files you have, although you might want to lock your audio tracks in the sequence so you don't accidentally change something while the mix is going on.

The second method is to make an OMFI Audio Only file. This creates an intermediate file that is actually another, more widely used audio format, AIFF, along with all the edit information. Once this very large file is moved over to the digital audio workstation, it can be converted back to an SD II file for use with the Macintosh Pro Tools software. The intermediate OMFI file is opened up in the OMF Tool that comes with Pro Tools and converted to a Pro Tools session. The OMFI file can be deleted after the conversion.

In order to send the audio mix back to the editing software, you must "bounce" the audio tracks to a continuous audio track. This real-time process changes the audio file, which now has subframe edits, to a frame rate that can be used by the Avid. Alternately, you can output to a digital tape format, redigitize it into the Avid, and line it up to the beginning of the sequence. Having some sort of synchronizing beep tones with a countdown or flashframe simplifies this final sync.

Here are the steps for moving audio from Mac editing systems to Mac AudioVision:

- Make sure you've digitized your audio using SDII media.
- Duplicate your final sequence and delete the video tracks.
- Consolidate the audio-only sequence to the transport medium (Jaz, Zip, Media Dock, whatever) using handles agreed upon by the picture editor and the sound designer.
- Export the sequence as an OMFI Composition of type AudioVision, *without* media.

Here are the steps for moving audio from Mac MC to Mac ProTools:

- Make sure you've digitized your audio using SDII media.
- Duplicate your final sequence and delete the video tracks.
- Consolidate the audio-only sequence to the transport medium (Jaz, Zip, Media Dock, whatever) using handles agreed upon by the picture editor and the sound designer.
- Export the sequence as an OMFI Composition of type SDII, *without* media. If using earlier than ProTools 4.0, use OMF 1.0; if using ProTools 4.0 or later, you can use OMF 2.0.
- Before using in ProTools, the OMF Tool must be used to convert the OMFI material into a ProTools session file.

Alternate steps for moving audio from Mac MC to Mac AudioVision:

- Duplicate your final sequence and delete the video tracks.
- Consolidate the audio-only sequence using handles agreed upon by the picture editor and the sound designer.
- Export the sequence as an OMFI Composition of type AIFF, *with* media. If using earlier than ProTools 4.0, use OMF 1.0; if using ProTools 4.0 or later, you can use OMF 2.0.
- Before using in ProTools, the OMF Tool must be used to convert the OMFI material into a ProTools session file.

This method converts all the media into AIFF and packs it into one file that contains both the composition information as well as the audio media.

The resulting file is very large, and due to the conversion required, export is much slower. Again, if you are going to a Symphony system or ProTools on Windows NT, consider using AIFF as the audio format throughout the production.

ADOBE AFTER EFFECTS

When preparing a composition in Adobe After Effects, you should always use certain settings when rendering for the highest quality:

- Always use 29.97 fps when exporting from Avid and rendering from After Effects (AE) in NTSC (*not 30 fps*). Of course, PAL is 25 fps and 24P is 24 fps.
- Graphics or other elements should be created at 720 × 540 (NTSC), or 768 × 576 (PAL). This is an accurate square pixel representation of the TV screen. Graphics are then resized in AE to the 720 × 486 (720 × 576) D1 pixel space.
- Whether one chooses to work in AE at 720 × 540, 720 × 486, or 648 × 486, it is vital that the final render takes place at 720 × 486. For PAL, the proper comp output is 720 × 576.
- If separating fields in AE, use upper field first if digitized on an ABVB system or a PAL Meridien system. If footage was digitized on an NTSC Meridien system, it should be interpreted lower field first.
- If field rendering in AE, choose upper field first if going to an ABVB system or a PAL Meridien system. If going to an NTSC Meridien system, it should be field rendered lower field first.
- If working in 24P, do not field render!
- Using the Avid Codec is currently the best conduit for going back and forth between Avid and AE.

If rendering a graphic for compositing in an Avid, there are two ways to deal with the alpha channel:

- If going to a Symphony or Meridien-based MC, rendering a QuickTime movie with an embedded alpha channel is fine, because batch import allows for more control of the files once they're in the system.
- If going to an ABVB system, I recommend that separate files are output for the RGB and alpha information.

This is easily accomplished in After Effects:

- Select the comp in the render queue and choose Add Output Module from the Composition menu.
- One output module should be set up to save the RGB channels at millions of colors with straight color (unmatted) at the AVR required.
- The other module should then be set up to save the alpha channel at millions of colors using the required AVR.
- After import, a matte key effect is applied to use the alpha as the matte channel. As described here, when providing Avid with a matte key and fill, make sure the RGB channels are using straight color, not premultiplied. Avid editing systems do not use premultiplied alpha channels yet.

Importing and exporting video, audio, and graphics has many variations, formats, and choices. With this flexibility comes complexity, so any production company should find the processes that work best for it, simplify them as much as possible, and be aware that you have many tools at your disposal.

7

Introduction to Effects

The effect capabilities of Avid editing systems are surprisingly deep. Even with the more limited layering features of the Avid Xpress, wonderfully complex and professional results can be achieved very quickly. At this stage in the technology, you can have two streams of video and an uncompressed title all play at once in real time (along with 8 channels of audio). With the Meridien video subsystem, you have all of this and real-time color effects as well. This means that color effects are not considered a separate real-time effect. You can apply a real-time effect to a color effect on a system using the Meridien video board and it will all play back in real time. The Media Composer offers from 8 to 24 tracks of video (depending on the model), and the Avid Xpress offers 8 as well. Symphony always has 24 video tracks. Even with fewer video tracks, you can nest, collapse, and mix down to extend and expand any system's potential.

With the cheaper hourly rates to create effects that are normally unattainable in the low- or mid-range budgets, some rendering seems quite an acceptable expense of time. For those productions that are used to paying $700 per hour, these effects are not going to truly compete. The last time I checked, time was still money. But what is the difference in price and the difference in value with the higher quality of a high-end compositing station? The lines between cheaper/slower and expensive/faster are blurring. Using hardware accelerators, like the Ultra Blue ICE board, and incorporating faster CPUs, it will soon be hard to ignore Avid effects as a viable possibility for all but the high-end needs.

There is much to be covered to deal completely with effects—too much for this book; however, some basics can get you past the beginner stage. There is nothing like the experience of a hands-on class, and Avid offers two of them: one for an introduction and one for more advanced features and techniques. This chapter can only hint at some of the techniques you will discover with enough time to experiment.

Anyone who is serious about effects should own the 3-D option. This not only dramatically expands the repertoire of tricks, but it also gives you more real-time effects. As you become more confident with effects, you will also become serious about nesting. Nesting is the feature that gives you more levels of video layering than you could ever practically use except for the densest of graphics sequences. This chapter will discuss nesting later after the basics. Nesting gives you incredible power with an extra level of complexity.

TYPES OF EFFECTS

There are three kinds of effects: real-time, non-real-time, and conditional real-time. Real-time effects can be handled by the compression and video boards without rendering as long as there are only two sources of video playing at the same time or there is only one effect at a time. If more streams of video or multiple, simultaneous effects are desired, then rendering is necessary. The exception to this is the ability to play real time and animate uncompressed titles over one layer of real-time effects. This just recently became possible with existing hardware with the release of versions 7.0/2.0. These titles are put into a downstream key (DSK) and are separate from the standard playback restrictions. The only drawback is that in order to scale them up or down in size, you must disable the DSK check box on the title in the Effects Mode.

Real-time effects have an orange dot over the effect icon when unrendered. The good news about real-time effects is that they can take advantage of a "fast render" scheme that allows them to render at between eight and six times real time. The faster CPUs cannot benefit from this faster rendering time because much of the rendering happens on Avid's special rendering boards. But a faster bus — the part of the computer that all information must squeeze through going to and from the boards — can make a significant difference. The PCI bus system (9500, 9600, and up) improves rendering times over a NuBus system (950, 8100, and AMP). Of course, Avid engineers are always trying to speed up rendering of effects, so the latest versions (running on the fastest CPUs) will usually render faster.

A non-real-time effect is too complex to be dealt with so quickly and sports a blue dot once it is in the sequence. It must be rendered by the Macintosh itself and may take 20 times real time or more to render. Clearly, if a real-time effect does the trick, it is preferable and your effect design should take this into consideration. An imported graphic with an alpha channel is a real-time matte key, but a matte key made by using the matte key effect in the Effects Palette is a non-real-time effect. This is because the imported matte key has a third stream, the alpha channel, which does not change over time. It can be loaded into RAM and cheated into being "a run length encoded" source, not a real video stream. A matte key created with the matte key effect from the Effects Palette assumes that the alpha channel will animate over time and so needs the full playback and computing power of a real video stream.

Certain effects are too complex and have too many moving, complex edges that must be anti-aliased for the fast render hardware to cope. Occasionally, these blue dot effects have an extra choice for rendering: draft quality or highest quality. They also have an extra icon that toggles to show either a stair step (aliased draft quality) or a smooth diagonal (anti-aliased highest quality). The difference between draft and highest quality for rendering time can easily double an already longer rendering time for this effect. So if you are working at low resolution for offline purposes or you are still changing the effect sequence on a regular basis, you will not want to spend the time on highest quality. You can change the quality when you render the individual effect, or when you "render in to out," draft, highest, or as set in each effect.

A conditional real-time effect is an effect that, under certain conditions, may switch from being real time to an effect that must be rendered. It is not a non-real-time effect, however, and the colored dot is green not blue. The green dot indicates that the conditional real-time effect can still take advantage of the fast rendering scheme. They are not nearly so slow to render as a true non-real-time effect. Without the 3-D option, the user must choose between real-time keys or real-time wipes. Whichever effect is not chosen is a conditional real-time effect. Real-time effects also become conditional when there are two effects on the same master clip. If you put a dissolve on a color effect, the dissolve becomes a conditional real-time effect. Defaulting to the transition effect for the conditional effect ensures that usually it is the shorter effect that needs to be rendered.

The one other time an effect becomes conditional is when two 3-D effects have completely different shapes; say one is a page turn and the other is a water drop, within five seconds or less of each other. One effect will need to be rendered.

The conditional status does not occur if two real-time effects are stacked on top of each other on multiple video tracks. They do not turn green, although something must be rendered. The system does not look vertically for effects, just horizontally, on the same video track. This is because the user may choose to monitor a track where these effects will be real time, so the status of the effect depends on which track has the track monitor icon.

When layering effects on top of each other vertically, the material on the top track always has priority. This is not a true multi-channel digital effect device in the way most people think of standard DVEs. It is more like having many single-channel devices. Each video track can be considered a separate channel of effects and, with nesting, much more than that. But it means that the separate video tracks do not interact with each other because they are each like separate sequences. This modularity allows quick exchanges of shots when it comes time to modify an effect. It also means that if moving objects are going to change their layering priority on the screen, then the lower object must be moved to a higher video track.

EFFECT DESIGN

Technically, AVR 75 does not support real-time effects on striped narrow drives. If you are not using versions 7.x/2.x or later, then AVR 77 does not support real-time effects at all, even on striped fast and wides. But people do it all the time. Every now and then they have a title over a complex background at AVR 75 or 77 and they get an underrun error. This happens when the drives and the computer cannot keep up a constant stream of video without dropping a frame, and the system stops playback and gives you this error. The error may say audio underrun or video underrun, but it is most likely a problem with the complexity of the video that is causing the audio to underrun (although simplifying the audio may fix the problem). You may have this issue with the new resolutions as well, so make sure the drives you digitize to are capable of playing back that specific resolution.

Good effect design tries to achieve the most spectacular effects with the simplest use of layers. The fewer layers, the fewer problems with trimming and rendering. Simpler design means most of the time you can modify faster because it is easier to figure out what is affecting what. If you can do something in fewer tracks, it looks better and renders faster.

RENDERING

Rendering effects can be significantly reduced by using some basic strategies. In general, only the top track must be rendered. Whenever you render an effect, you are rendering a composite of everything below. Unless you want to play the tracks below by themselves, by moving the video monitor down to lower tracks or stripping off the very top tracks for multiple versions, you can leave these lower tracks unrendered.

What confuses people is that many times there is not just one track available to render as the top track. The beginning of the show may have a complicated layering section that has ten layers. Then most of the show may not go above track

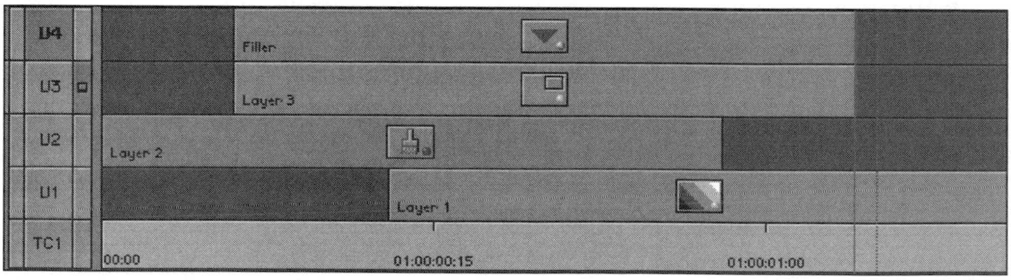

Figure 7.1   Applying a Submaster

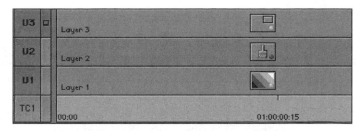

Figure 7.2 Rendering the Top Track

three and the end has five tracks. How can you set it up to just render a top track and walk away? The secret with older versions is using a fake effect—the submaster—and putting it on the video track above the effect sequence, just to have something to render. The newest solution is ExpertRender™.

To make sure that the submaster goes on the absolute top track, when you create a new video track, hold down the Alt/Option key. Then you are given a choice about which track you actually want to create. If you create video track 15 and place your submaster there, it is likely that you will not really need all of the tracks below. If you continue to add video tracks, then they are added incrementally, track three, then track four, etc., and leave your top track unaffected. Highlight the top track that contains the submasters, mark inpoints and outpoints around the entire sequence, and render in to out.

A more complex approach is to render with a submaster and then plan to leave a top track in real time if you can. The titles always look slightly better in real time (they stay a high-quality DSK), and you can fade them in and out with keyframes instead of dissolves. A new Fade Effect button in later versions adds keyframes for fade-in and fade-out to imported matte keys as well as titles. Keeping a Picture-in-Picture (PIP) or 3-D Warp in real time on the top track means that you can continue to make changes right up to the last minute while the lower layers stay rendered. Those last-minute changes to the top layer effect may be the change of the shot for something else, for a second version of the effect, or just a final tweak of the move itself.

Here's a new trick for adding a submaster to a track below the top video track. You can use this with 3.0/9.0/2.0 and later. Hold down the Alt key when creating a new track like described above. You will be given a choice of exactly which track number you would like to create. Choose the number of the top track and the system will tell you that track already exists. You can now choose to insert the new track and push up the existing track. For instance, if your real-time track, the one with the DSK titles and the DVE you don't want to render, is track seven, then it becomes track eight and the new track becomes track seven. You can then add your submaster to track seven and render, or you may want to experiment with ExpertRender.

ExpertRender

ExpertRender is a new feature designed to let the intelligence of the system solve your rendering problems for you. If the main problem with rendering is that people render too much, then the best solution would be to make sure everyone feels comfortable with a minimal style of rendering. The reason people render too much is that they don't really know what will play in real time and what won't. This is because many effects are conditional and depend on what else is going on to determine whether they are real time or not. Don't take the time to step through a long and complicated sequence effect by effect and still, perhaps, guess wrong. It is easier to mark in at the beginning of the sequence, out at the end, turn on all the video and audio tracks, and use render in to out.

Rendering in to out may be easy, but there are some definite drawbacks. The first is the extra time involved in rendering effects that will play just fine right now. The other problem is that when you render graphics or titles that are using the DSK (downstream key), you are taking an uncompressed image and bringing it down to the level of the rest of the sequence. Of course, if you are working at 1:1 resolution, then this will not create a noticeable difference, but at 3:1 or 2:1 you may be able to see the difference with complicated images. Lastly, any time you want to go in to change any of the effects that could have been real time, you are unrendering them when you make the changes. You may think that you are all done and ready to go to tape, but your client may have other plans, and you want to stay flexible to the very last minute.

The system knows what it can play and what it cannot. When preparing to play a sequence, the system must "build the pipes" and allocate the hardware resources on an effect-by-effect basis. The Avid player must look at each effect and say, "I have a video stream available for this effect and a DSK free for this effect, but this one here must be rendered." This all happens at a very deep layer of the programming code and is very basic to the functioning of the machine, so it is very dependable. We can reveal this capability to the user in the form of ExpertRender.

By marking in and out, turning on all the video tracks, and choosing Expert-Render in to out, we tell the system to do an analysis before rendering. The results are then displayed in the form of a dialog box that says "X effects out of Y will be rendered," and it will highlight those effects in the Timeline. You can now see the direct benefit of how many effects you do not need to render. They will now stay real time with a guarantee that they will play. The results can be astounding with only a handful of effects out of hundreds really needing to be rendered.

There are times when you may disagree with the Expert. Specifically, you may have plans for a certain section and you will be adding more effects to a higher track when you are done with the rendering. In this case, the system cannot read your mind to know what you will do next and can only return results based on the existing sequence. You can then choose to Modify the ExpertRender choices. By clicking on Modify in the ExpertRender dialog, the Expert leaves all of the chosen effects still highlighted in the sequence. You can Shift-select or Shift-deselect as you see fit and hit the regular Render button when you are done. There is no need for a render in to out again because all the effects are already selected.

Another time you may disagree with the Expert is when you have dissolves between titles. This is relatively rare and should really be treated as an exception. In this case, the system will realize that the real-time dissolve cannot be played in real time, but because of the order that it must allocate resources, chooses the titles for rendering. Again, the user can easily override this situation, pick the shorter answer, and render the dissolve only.

The beauty of this automated analysis is that a vast majority of the time the choices are the shortest rendering answer. In reality, you will save so much time by letting the Expert do the job for you that even the occasional overrender is easily overlooked. How much time is wasted stepping through effects by hand? The guarantee that the system will be able to play the entire sequence after the ExpertRender is, all by itself, money in the bank.

Partial Render

Partial Render is the new ability for the system to render only part of an effect at a time and then come back later and pick up where it left off. This allows you to start a render at any time, even if you know you don't have enough time to finish rendering the entire effect. By hitting Ctl/Cmd-period you can escape from the render and keep or discard what has been rendered so far. This is especially useful if you have a series of slow, blue dot effects and can start to render a little more anytime you take a break.

The system will create a new precompute for each partial render and tie them all together to play the final effect. You can see how much of an effect you need to render by changing the Render Range in the Timeline view to show "Partial." This is the most useful setting and should be left on most of the time since it will show you only what is left of an effect that has started rendering, but not quite finished. If you change the view to "All," then it will show you all effects in the sequence that are not rendered. Although this can be useful under some conditions, it can be confusing if you are relying on ExpertRender to figure out what needs to be rendered. The visual feedback in the Timeline of the red or partial red line across the top of the clip with the effects can clash with the information from dupe detection. I would strongly suggest that Render Range display and Dupe Detection not be turned on at the same time. If these two functions are important to you, then they can be made part of a Workspace and changed with a single keystroke.

The only other drawback to Partial Render is that with long-term complicated effects projects you will be generating more precomputes. If you have been cleaning up precomputes once a week to keep the system operating without the high, unnecessary overhead of too many small files, you may want to do it more often. Most of the time, however, you won't even be aware that Partial Render is at work. The best features, many times, are the ones that make you more productive without attracting attention to themselves. Long after the sizzle of the product demo is over, you will be making your deadlines with projects you are proud of, and you won't really care why!

SOFTENING

As layers are added to a sequence and there is more detail in the image, some extra compression of the image is required to ensure playback at that AVR using the ABVB board. This is the drawback of working with lossy compression and is the main reason many people spend the big bucks for uncompressed effects. Minor amounts of compression are handled invisibly, but when the level of complexity gets too great, you need to soften or simplify the effect.

You may get a message that an effect must be softened while it is rendering. If at all possible, never soften an effect on purpose. Not only is it unacceptable for finished quality work, but it also drops out of the Fast Render Mode to do the softening. I always set the Render Setting to "Do not soften." This means that at a later date I might get an underrun error message because the drives cannot keep up with playing back the level of complexity of that image. If I get the underrun error, I do almost anything else before softening. You can reduce the amount of audio tracks with an audio mixdown or render all the real-time effects around a trouble spot to help simplify that part of the sequence.

The softening message you might get on ABVB systems during rendering is a little conservative. Avid engineers are always planning for the worst-case

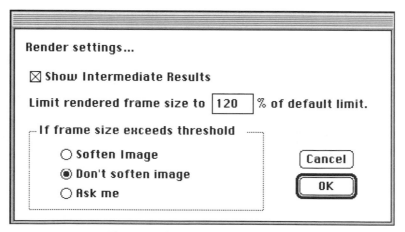

Figure 7.3 ABVB Render Setting

scenario of a maximum amount of audio tracks, rubberbanding audio, and highly complex real-time effects when estimating the ability to play back a complex image. You may never get near that level of demand on your system in normal work. This means that if your sequence is simple, but the individual effect is complex, you have some wiggle room to push the machine beyond the prescribed limit for a short time. Some people like to know if their effect is too complex. They may have problems playing back later, so they leave the Render Setting button at "Ask Me." Then they set the render limit number above 100 percent to maybe 120 percent. This means they are only alerted if the effect is seriously over the supported limit of complexity. I always put this setting to "Don't soften" and have almost never had a playback problem because of effects; however, if you have non-striped, slower drives, this may occasionally cause a problem with high AVRs.

KEYFRAMES

Almost all effects can be manipulated by keyframes. The only exceptions are the color effect (on non-Meridien systems) and a few other segment effects like flip and flop. Keyframes are the method to change an effect over time, so there are always at least two keyframes: beginning and end. At the first keyframe, certain values about position, shape, or color are entered, and the settings change to match the values on the next keyframe. The change, if there is any, is smoothly interpolated between the keyframes. Keyframes can be added in the Effects Mode on the fly while playing and hitting the Keyframe key (the Apostrophe key) or, on older versions, by mapping the Locator button to a Function key (this doesn't work in the Wireframe Mode with 3-D). Keyframes can be copied and pasted, dragged by holding down the Alt/Option key, and, in version 6.5 and later, moved with the Trim keys. If you want the effect to just hang on the screen, with no motion, then you can copy and paste the same settings between two keyframes or highlight both keyframes when changing parameters.

Keyframes and Speed

The speed of any effect depends on the time between the keyframes and the amount of distance the effect must travel on the screen. If the keyframes are close together in time and the distance on the screen is great, then the effect appears to be fast. The overall length of the clip in the sequence determines the distance between the first and last keyframes. If the effect is put onto a very long clip, it takes the length of the clip for the effect to end.

Many times there is a situation where you want the effect to move rapidly and then stop with a longer duration clip. This requires three keyframes. The first two will be close together in time to provide the proper speed of the move. Then there is no change between the second and third keyframes as the shot plays out. The only exception to this is when using the Spline parameter with the 3-D Effects option. Spline always tries to smooth the path between keyframes—even if they are not moving. With Media Composer and Symphony, you can turn Tension up to 100 in the Spline parameters so that there appears to be no movement between two keyframes that do not otherwise change position.

If you trim a clip and make it longer after adding an effect, the effect gets longer, too. If you trim a clip with an effect shorter, you may create an impossible situation. If there is a keyframe on every frame and then you shorten the effect, where do the keyframes go? No one knows. You need to adjust this effect by hand after shortening because the results will be unpredictable.

SAVING EFFECT TEMPLATES

After all this work making the effects just right, you can save them as effects templates so they can be easily applied over and over again. Any effect can be stored in a bin by clicking on the effect icon in the upper left-hand corner of the Effect Mode Palette, and dragging and dropping the effect icon in the bin. If a bin that holds an effect template is open, that effect will be available in the Effect Palette. The effect template can also be dragged back from the bin and applied to the Timeline. When you apply the effect, it looks slightly different from the original effect unless the new clip you apply it to is exactly the same length as the original clip.

If the Alt/Option key is held down when dragging the effect icon to a bin, then the effect template is saved with the video. This is an "effect with source" and can be edited into the sequence like a master clip. If you add an effect to a title, then just saving the effect template for the title is always "with source," so you don't need to hold down a Modifier key. If you want the title effect template to be just the keyframes alone so the effect can be applied to another title, then hold down the Alt/Option key when saving it. Here's how to remember it:

- Alt/Option drag the effect template for effects gives you an effect and the source clip. Use this like a subclip with an effect attached.
- Alt/Option drag the title template on titles gives you the title's keyframes and no source. Apply this template to another title to get a similar title move.

ADD EDITS

If an effect cannot be manipulated with keyframes, it can be split into sections using the Add Edit button. By splitting one effect into multiple effects, each one can be manipulated separately and then recombined by a dissolve. This is most useful for a color effect that can be used to change a color over time with add edits and dissolves. Due to hardware restrictions on ABVB systems, the color effect may occasionally have a flashframe at the end if the effect is a radical change from the normal master clip. This flashframe can be fixed by rendering the entire color effect. The Meridien systems will play both color effects and dissolves in real time on the same clips.

Creating add edits adds extra keyframes to an effect sequence. If you have an effect with two keyframes and you split it with an add edit, then you have two effects with two keyframes each. If the original effect had a smooth motion across the screen and now it has double the keyframes, you could have a problem with acceleration. Acceleration is an effect parameter that smoothes the motion of an object's path across the screen with a slow-fast-slow speed change. Acceleration is turned on many times to approximate the physics of natural motion. Adding an extra keyframe causes the effect to slow to a stop at the new point and pause where there used to be a continuous motion. If your effect has an object in motion with acceleration, you should apply only an add edit at a point where a keyframe already exists.

NESTING

Nesting is the most complex, but the most powerful, characteristic of effects with the Avid editing systems. So far we have discussed building effects vertically, which, depending on your model system, may be limited by the number of tracks. Once you understand nesting, you can expand the amount of tracks dramatically. Nesting involves stepping into an effect and adding video tracks inside. More effects can be added inside the nest, and then you can step into those. It is a way of layering multiple effects on a single clip, but also much more. The only real limitations are how long you want to render and how much RAM you have.

There are two methods to view a nest with Media Composer/Symphony and one with the Avid Xpress. With Media Composer/Symphony you can apply an effect and then use the two arrows at the bottom left of the Timeline to step in or out (these buttons are mappable to the keyboard in later versions). Once inside the nest you can no longer hear audio, but you can focus on that level alone and work on it like it is a separate sequence. Within that layer you can add as many new video tracks as your model allows. Red numbers on the timecode track in the Timeline will indicate how many layers you are nested in later versions.

Figure 7.4 The Up and Down Arrows for Traditional Nesting

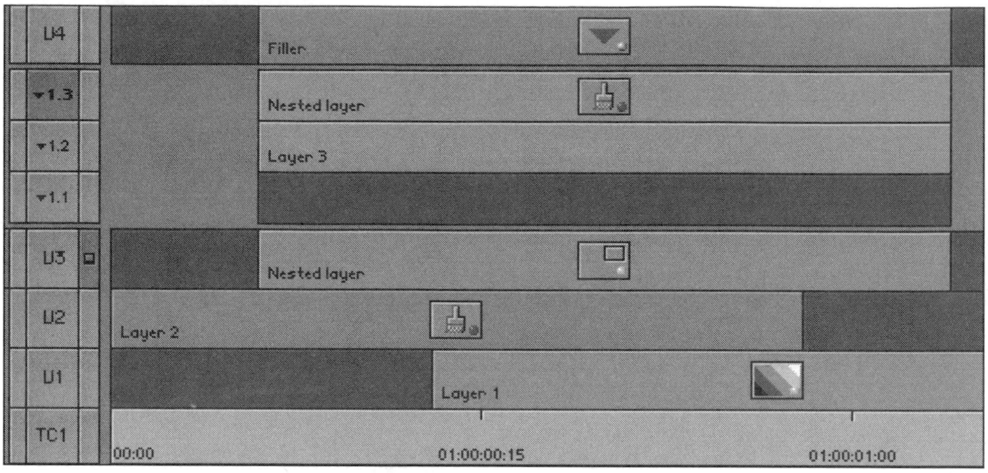

Figure 7.5 The New Nesting Display

The other method to view nesting is used on all models. Using the segment arrow to double-click on a segment with an effect, the tracks in the Timeline expand to see all the layers inside at the next level. Continuing to double-click layers inside the first effect reveals those tracks as well. Tracks can be patched and edited in this mode, and the audio can still be monitored. It is a little easier to understand all the effects going on because the display is more graphic.

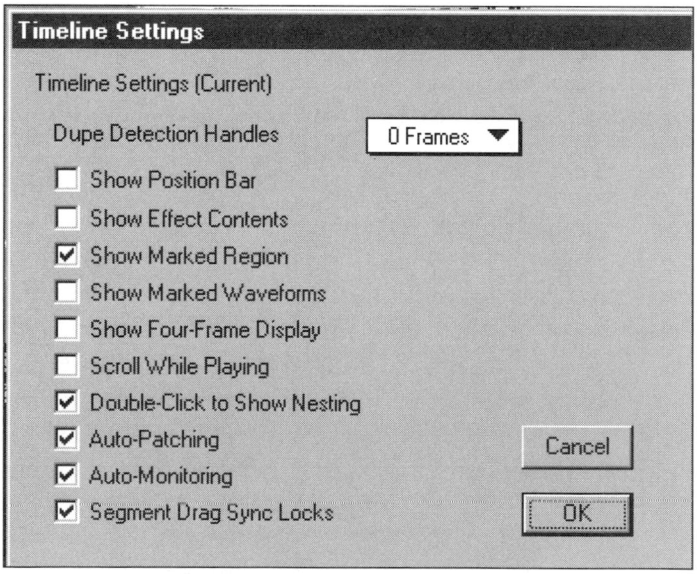

Figure 7.6 Timeline Settings

This mode of viewing nesting can frustrate people who open it up by accident and then are confused about what they are looking at. You can always close the expanded view by double-clicking on the original track again with the segment arrow or, in Media Composer or Symphony, Alt/Option-clicking on the down nesting arrow. The view can be turned off in Media Composer and Symphony by unchecking the checkbox in the Timeline Settings called Double Click Shows Nesting. I recommend turning this feature off if you like to move very quickly and you have an older Mac.

Autonesting

Nesting this way implies a certain order of assembly: Apply the outer effect and then step inside. You must apply the PIP and then step in for the color effect. But real life doesn't always work this way. Many times the nest is a secondary thought, used well after the first effect is in place and rendered. In this case there is autonesting:

- Select the clip with the segment arrow in the Timeline.
- Alt/Option-double-click an effect in the Effect Palette.
- The second effect does not replace the first effect, but it covers it.

This adds the layers from the outside instead of stepping into the effect and building them from the inside.

All of these methods are for adding multiple effects to a single clip, but they are just as useful for adding one effect to multiple clips. If there is to be a color effect that covers an entire montage, then it is a waste of time and energy to put a separate effect on every clip. What happens when the effect must be changed? Now you need to change one, turn it into a template, and apply it to all the others. But there is a faster way that uses autonesting:

- Shift-select multiple clips in the sequence with the segment arrow.
- Alt/Option-double-click on the effect in the Effect Palette.
- The effect autonests as one effect that covers all the clips.
- Adjust the one effect and all the clips are changed.

If you want to change one of the clips and replace it with another shot, just step inside the effect and make the edit. You can also step inside the multiple-clip effect and add dissolves or other transition effects. You must render these inside effects, but you can leave the outside effect in real time to allow for future changes. Of course, if you render the outside effect, it will create a composite of everything inside.

Figure 7.7 Autonesting

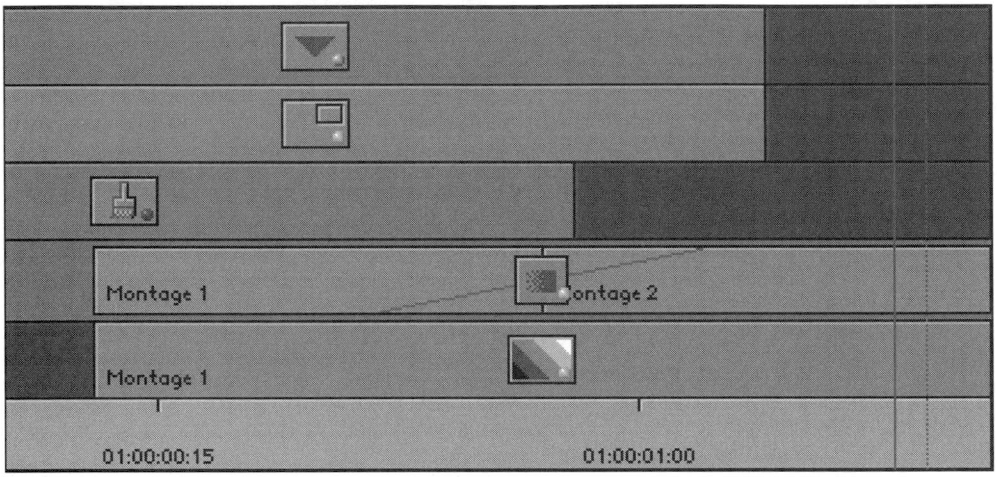

Figure 7.8 Displaying a Nest

Collapsing Effects

You can nest an entire effect sequence into one effect after it has been built in a more traditional fashion using vertical layers. If keeping track of all the video layers becomes tedious, you can collapse them into a single layer. In order to nest effects, you must have an outside effect at the outermost level. Collapsing places a submaster effect over all the layers and nests them inside. Selecting the area to be collapsed by marking in and out and highlighting the desired tracks, you can collapse all the tracks except the top track, which, like a title for instance, you can leave in real time.

There is no way to really uncollapse an effect segment. Here is the best method:

- Step inside the collapsed effect.
- Mark an inpoint and outpoint around the entire segment.
- Turn on all the video tracks (except for V1 if it is empty).

- Use Alt/Option and the Clipboard button while inside the Collapse.
- The layered segment goes directly to the Source window to be used as a subsequence.
- Cut the subsequence back over the top of the collapsed effect in the Timeline or drag to bin.

Rendering a nested effect is simple; just render the top, outside effect, the submaster. This leaves all the effects inside unrendered, but it is sufficient to play as long as you are monitoring that outside effect.

You don't need to leave the top track of a collapse as a submaster. You can replace the submaster with another segment effect by just dragging and dropping from the Effect Palette. You can replace the submaster with a mask or color effect. Again, just like the strategy of leaving the top track in real time in a traditional, multi-track effect, if you step into a nest and render the top track inside the nest, the outside effect can continue to be manipulated. You can keep tweaking the effect on the outside of the nest if the dissolves inside are rendered.

Collapse vs. Video Mixdown

Although the collapse feature is excellent for simplifying complex effects sequences down to one video track, a collapse can still potentially become unrendered. If you are sure that an effect sequence will never need to be changed, match framed back to an original source, or used for an EDL, then you can use video mixdown. Video mixdown (under the Special menu) takes any section of video between marked points, whether it has effects on it or not, and turns it into one new media file. This new media file has no timecode (which timecode would it use if you had fifteen layers?) and breaks all links to the original media. This is why match frame and EDLs are no good.

Video mixdown should be used only for finishing and for something that you will be using as a single unit over and over, like the graphics bed for an

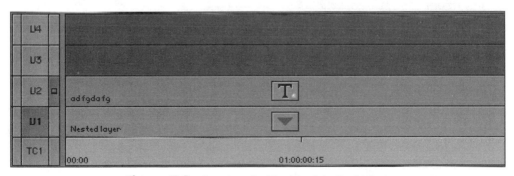

Figure 7.9 Leaving the Top Track in Real Time

opening sequence you use every week. Once all the effects are rendered, then a video mixdown is as fast as copying the media and the rendered precomputes to another place on the drive, an insignificant amount of time. If the effects are not rendered before the mixdown, then they will be rendered first as part of the video mixdown process, but don't forget to count on the rendering time in your calculations. So the workflow encourages rendering first and video mixdown later when everything is signed off.

A video mixdown will significantly improve the performance of the Avid during long sequences with lots of effects. Instead of forcing the computer to "build the pipes" for lots of complex effects with many short media files, it just needs to find one master clip. This means snappier reaction time when you press play. Always make a copy of the sequence before you overwrite a mixdown over your timecoded original sources so when the client changes their mind you will have a fall-back sequence. Video mixdowns are very powerful and time-saving for a wide range of purposes, but don't use them for offline if you plan to redigitize or make an EDL!

3-D EFFECTS

In Media Composer and Symphony, all 3-D effects come from one effect, the 3-D Warp, or you can "promote" titles and imported matte keys. In Xpress there are some premade effects broken out into simple-to-apply subsets of the entire 3-D functionality. Titles and imported matte keys can be animated around the screen in real time using the Promote button. Having the extra piece of 3-D hardware gives you both real-time wipes and real-time keys.

The 3-D Effects option and "m" AVRs are completely incompatible because to enable four screens to play at once for hardware multicam, the "m" resolutions are at a much reduced frame size. The 3-D Effects option must be disabled before working with this unusual frame size. This is accomplished by launching the application and holding down the "F" and "X" keys. You will be given the option to enable or disable the 3-D.

There are clearly more things you can do with the 3-D effects than just spin and tilt. You also have Stamp and Clear, which uses a real-time buffer to stack unlimited layers without any rendering or loss of quality. These stamped layers must be added while the sequence plays in real time; they cannot be rendered and must fade out all at once, but the possibility for titles is very powerful. Remember to clear a stamp buffer or the effect will pop back onto the screen after you have faded it out. And the ever-popular trail effect can be added to any object that moves by taking advantage of the same buffer used for Stamp and Clear. Trails also are real time only; render the trail and it disappears.

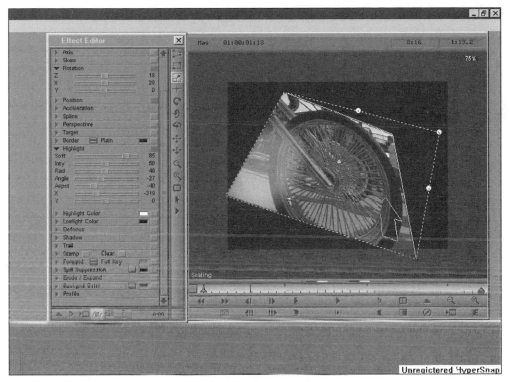

Figure 7.10 3-D Effect

There is also corner pinning, which allows you to fit four corners of an object so that it matches the edges of another object. It is not quite morphing, but it allows you to put images inside TV monitors, picture frames, or the like. The animator's tool for realistic motion, Spline, is also useful for getting fast, professional results for complex motion paths.

There are so many things you can do with the 3-D option that I helped write a one-day course just for that. Truly, this area calls for personal experimentation. Just taking some of the shapes and using them to warp images into interesting moving backgrounds requires parameters you must discover for yourself.

PAINT AND ANIMATTE

Version 7.0 includes significant additions to the capabilities of intraframe effects. If you consider editing to be an interframe process — working between frames — then painting on a frame is intraframe. There is a full palette of familiar choices for anyone who has used third-party painting programs. Brushes can be changed and areas of the image can be blurred, color corrected, and generally affected like any standard paint program. Multiple layers of paint effects on the same frame are possible, but unlike the other, single-image paint programs, you can easily change the effects over time with keyframes.

You save significant time and trouble by keeping the entire process inside the single editing application. No exporting, painting on separate frames, and re-importing anymore. Although some programs, like Fractal Design's Painter, allow overlapping of multiple frames with an onion-skin-type interface to reproduce what cell animators use, you cannot say, "Start on this frame and expand the effect until this frame." Interpolation between keyframes is the beauty of working in a video-based program for paint.

You can also isolate parts of the frame, say the sky, and draw a matte shape around it to make it a deeper blue. The ability to draw on an object and create control points to adjust curves with bezier handles and move the edge over time means that almost any part of an image can be manipulated separately from the whole. By creating points that change over time, any smooth, even motion in a shot can easily be followed with just a few keyframes. If the motion is jerky or unpredictable, you need more keyframes to adjust the control points. If the motion is truly difficult to follow or you are doing dozens of motion-following effects, you may want to use another program that allows you to automatically track specific pixels over time. Puffin's Commotion or Eyeon's Digital Fusion are both good choices and both read OMFI media files.

The Color Effect has been improved to allow quick adjustment of contrast range and gamma using eyedroppers. Pick the darkest point of an image with the eyedropper and it changes to the blackest possible values. Choose the brightest

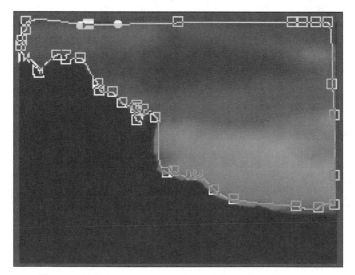

Figure 7.11   Creating Control Points on an Object

part of the frame and that will become the brightest possible value for broadcast. An image that was dull because of a narrow contrast range can make use of the full range of levels available. You can also quickly take illegal levels and bring them to within broadcast specifications for luminance and black level using the adjustable clip levels. With the latest versions, there is also the Spot Color Effect, which is the combination of all the different choices in the Color Effect with Intraframe. It also includes NaturalMatch™, which is a powerful color-matching algorithm from the Advanced Color Correction Mode. This could be considered a keyframeable secondary color correction.

AVX

A whole new range of choices is now available with the creation of the AVX (Avid Visual eXchange) plug-ins standard. This is an interchange format so that other third-party effects companies can easily modify their existing product line for use as an Avid plug-in. These video-based effects are much better suited to use in a video-editing program than effects created specifically for a paint program.

We have just begun to see some of the fantastic effects coming from third-party AVX plug-in creators. There are interesting packages from Hollywood FX and Avid's own Transjammer. Cinelook from DigiEffects gives an amazing amount of control over turning video into a "film look." Boris Effects creates an entire software DVE and titling tool for those who need the extra control and creativity. One already very popular AVX plug-in is the Ultimatte plug-in. The amount of control over keying with colored backgrounds and the excellent edges with basic default settings means you can finish special effects projects shot over blue or green screen on the Avid.

Many of these AVX effects have been "ICEd," which means that ICE (Integrated Computing Engines) has modified these effects to take advantage of their multi-processor rendering board. The folks at ICE take existing AVX effects (and plug-ins for other third-party programs, too) and change them to render in multiple parallel pieces. Depending on the effect, this can make the difference between something that is nice to have and something you use every day. Most Avid editing systems have now been qualified to work with the ICE board, and some resellers will not sell a Symphony without one. Just make sure you have the latest fan kit for your CPU before you add in such a large and (ironically) heat-generating board.

TITLES

Higher quality titles are available in versions 7.x/2.x and later with the uncompressed choice for titles playing back in the DSK (downstream key).

Aliased edges and blockiness are eliminated when titles are uncompressed. There is also the ability to run an uncompressed title on V2 and to have unrendered real-time effects on V1. This is because you are not running three streams of video, just two streams and a PICT file. With the ability to copy and paste from a word processing program, instantly apply a custom style, and create title rolls, it is becoming faster, simpler, and more feasible to use the Title Tool for large amounts of text. With later versions you can create a custom title template and map it to a Function key. All styles are automatically mapped to the next unused Function key and are only enabled when the Title Tool window is active. If you are creating titles with lots of font and size changes, you can highlight the text in the Title Tool and hit the Function key to apply the premade style.

You can use the Paint functions of Intraframe to blur titles with the editing application using this very basic trick. You can use a variation of this technique to put any kind of effect on a title:

- Cut the title into the sequence.
- Step into the title and mark the entire length of the title fill on V2.
- Copy this to the clipboard. Alt/Option-Copy to Clipboard saves a step later by automatically putting the fill in the source window.
- Step out of the title and use Remove effect. This will remove the fill and unlock the alpha channel.
- Add a new video track above where you are now and move the title alpha to the higher track using the segment arrow.
- Cut the fill back underneath the alpha channel onto the track below.
- Apply a Paint Effect to the alpha channel. Select the entire frame and apply a blur. You can continue to manipulate the blur using keyframes.
- With the alpha channel still highlighted, get the Matte key effect from the Effect Palette and Alt/Option-drag it over the Paint Effect. This will autonest the matte on top of the Paint Effect.
- Render the effect.

For truly complex manipulation of type, you still want to work with a program that can use vector-based graphics. Softimage Marquee now ships as a standard AVX plug-in on Symphony and is an option on other NT systems. This true 3-D type and graphics manipulation program allows you to quickly create titles with textures, light sources, and extruded type. You can manipulate each letter in a title on its own Timeline and control all the movement with bezier curves. A static title is quick to create and plays back in real time. An animated title will take longer to render, and you may want to consider using a stand-alone workstation for that rendering work. The title can be previsualized over the video in the editing program, then the file can be sent to the rendering station to use the power of an Open GL board. Once Avid systems can use Windows 2000, they will also be able to take advantage of an Open GL graphics rendering acceleration board. Because of the existing restrictions on monitor configurations on Windows NT 4.0, Avid cannot use these boards just yet.

The ability to layer, paint, and use plug-ins has given the Avid editor a whole range of tools and looks that were never available before in one package. It also makes certain effects, which were out of range for small-budget programs, easily accessible. The trend in effects these days is slowly moving away from the "Wow!" to more subtle, invisible applications. But there is still a great need for the ability to turn any piece of video into a graphic or give it a graphic look. Faster rendering, acceleration boards, and more real-time streams are coming for nonlinear editing systems. Faster CPUs promise that more work can be done without dedicated hardware. Networks will allow users to distribute rendering to unused or dedicated rendering systems. Editors and designers will always continue to experiment and push the technology to the limits and, with the tools now becoming available, nonlinear editing with the Avid will continue to generate new and fascinating high-quality effects.

8

Preparing for Linear Online

Most of the preparation for a linear tape-based online happens at the end of a nonlinear project when you make the edit decision list (EDL). But there are things that must be done properly throughout the job in order for the online to go smoothly. The reality is that with all the twisted financing that goes on in this crazy industry, there are still reasons to buy a bunch of offline models and create an EDL when finished. For those who have bought the lower end models, which do not have the finishing quality resolutions, you need to prepare for a linear online edit session at the end of every project.

This chapter will discuss the most basic requirements to get a good EDL under the most common scenarios. Then we'll go a bit deeper into the possibilities for increasing the speed of an online auto-assembly, the Holy Grail of all expensive online sessions, and almost as tough to find! But the good news is that many facilities want the list in a simple, bulletproof form so that they can do their own, more complicated variations.

How do you transfer a 24-layer effects sequence with nesting, color effects, audio equalization, and rubberbanding to a format that supports one layer of video, four tracks of audio, and hasn't changed much in almost 20 years? Answer: Not very well. Dedicated hardware is the rule in traditional online suites with a different piece of equipment and, many times, a different manufacturer for each function, like text, special effects, and tape control. Many times the separate pieces of equipment were purchased years apart and there may be components that are ten years old that still work fine for what they were designed to do. Compare the processor of a CMX 3600 with a Macintosh G4 or IBM Intellistation. The Macintosh and IBM have sped up their processors by over 100 megahertz per year and show no signs of slowing down. Yes, if you design the hardware to do just one thing, then you can maximize it so that it doesn't need as much RAM or processor speed. But just try to add new features! This is why there are so many third-party EDL management software choices. If the dedicated linear online systems cannot add new software features, then new third-party software does it and simplifies the EDL to a format the dedicated systems can handle.

Even with third-party EDL programs, there is still a massive mismatch in capabilities between what can be done by clever folks inside the Avid and what can be represented by this limited format for a linear online. The only way to convert from one format to the other, preparing for that precious linear online time, is to dumb everything down. You need to understand the limitations of what can be conveyed in an EDL. Bring the Avid EDL to a point where the show can be built back up again, piece by piece, in the order necessary in the linear, tape-based world.

This is more of a challenge than it would appear at first, not because it is that difficult, but because of the limited experience of the Avid editor in the world of linear online. Unless you have actually worked as an editor both at an Avid system and at that particular linear online suite, you are not really in a good position to take advantage of all of that tape suite's advantages. We can discuss capabilities in generic terms, but that doesn't help you for the specifics of that edit suite. You need to know that for that suite, you needed to reserve the character generator a day in advance and you can forget using the digital effects unless you book a night session.

This is where you must rely on two very important low-tech tools: clear communication and the ability to collaborate with experts. As you devolve the Avid list to something simple for the linear assembly, you must be very clear about exactly what must be done to every part of your sequence. What effect do you really want? How are you going to deal with all those channels of audio? This chapter will describe different ways to leave a paper trail that someone else can follow. Count on the expertise of the operations or scheduling people at the particular post-production house that has been chosen. Speak to the editors if they are available. Depending on time available and the complexity of the program, you may even want to (gasp) plan a pre–post-production meeting to discuss approaches with the editor! Imagine, warning them what to prepare for! Let's face it, there is no way to predict everything that can happen after you leave a project. It is really not fair for the offline editor to be blamed for changes that occur after the cut is locked. Sometimes changes ripple backward to affect a decision you made in good faith, with the information available weeks or months ago. Unfortunately, this happens all the time.

Coming from a post-production facility background, but having freelanced for a few years as well, I strongly advocate building a good relationship with a handful of production companies. Bring any work you can to these select production companies, large projects and small, and spend some time learning the facility's capabilities and quirks. If you create a sense of loyalty, they will look out for you. This happens in little ways, like working a little harder to fit you in for emergencies. They may come to know your work and style and be able to anticipate and fix problems for you before you even know they are there. If they can get you in and out on time and on budget, they can fit in more paying sessions, need fewer "make goods" for questionable mistakes, and lower their already unnecessarily high stress level.

If you can convince your employers to pay for you to sit in on a few online sessions, it will pay off for them in the long run. If they have a relationship with you, then it is in their best interest for you to continue to make informed decisions about effects that can be re-created easily, or faster ways to assemble the show in the final stage. Will it be faster to dub shots onto a selects reel? Is it better to have more tape decks available for a shorter time or fewer decks for a longer time? And really, what can you do at a lower rate per hour to make things go smoother? When you learn these things, you become more valuable to your employer and have more job security. You are more likely to stay on the "A List" the next time your favorite client, director, or producer has a big job.

So it is in everyone's best interest that you understand the implications of every step of your offline on the final online. If you digitize without timecode or without paying attention to tape names, you are directly hindering the next step. Don't blame "those Avid EDLs" when you have no regard (and, really, no respect) for the next link in the chain. It is your job to continue to inform the director or producer that choices they are forcing you to make will cost them money later. If they take just a little more time now, at the lower rates, things will go smoother at the crunch time.

What are the biggest mistakes? Ignoring the importance of timecode and tape name. Everything must have a timecode reference—all sound effects and music, all graphics and animations, and all picture and sync sound. Let me say this again: If you are going to a linear online, every source must have timecode. It is tempting to just import that one cut from the CD music library or just pop in those three shots from the VHS dub. But you must be sure to go back and match by eye or ear with a timecoded original source before you make the EDL. This is a clear example of GIGO—garbage in (sources without timecode) and garbage out (a list that cannot really be assembled).

PREPARING SOUND

There is one exception to the timecode rule: when the audio digitized into the Avid will actually be used as a source in the online (or the final mix for a film). It is becoming more common for the first edit to lay down a digital cut of the audio, low-resolution video and then cut with original video sources the rest of the day. This assumes no changes in the online that will affect sync or length, of course, something you may have to mention repeatedly! In this case, you are working from the timecode of the digital cut, which should match the timecode of the master sequence. If the show starts at 1:00:00:00, then the sound source should, too. Editing the high-quality digital sound works especially well with timecode-controlled DAT or another digital tape format like the 8-channel DA-98. You could go out of the Avid using the digital audio output format to a DigiBeta tape. In a pinch, audio tracks 3 and 4 of a Betacam SP can be used since these audio tracks have a higher dynamic range and lower signal-to-noise ratio than tracks 1 and 2. This is good for effects and sound under but should not be the main method for the most important sound in the project. Just make sure the Betacam SP decks in the online suite are models that can play tracks 3 and 4. Generally, this means a Sony editing deck with the letters BVW in front of the model number. The non-standardization of high-end Betacam SP decks makes this problem common in lower cost suites. Did I mention you can't make changes?

USING THE OFFLINE CUT IN ONLINE

If you are using digital videotape to bring the Avid digital cut to the online suite for your audio, then you should also include the video as part of that first edit. Many online editors use a clever trick where they lay down both video and audio from the Avid to the master tape in the online suite. They put a big circle wipe with color bars in the center of the video. That way they can cut over the top of the old video all day and see if there is any discrepancy between the EDL and the digital cut. Any time they see a flash of color bars they know something is a frame or a field off and can make adjustments. It sure is better than squinting all day to tell the difference between one frame of 10:1 and the original!

The other way that the digitized audio without timecode from the Avid can be used is when you make an OMFI export of the audio and the sequence for use on a digital audio workstation. But if the original sound was digitized with a little distortion or mono instead of stereo, then the sound mixers have to redigitize it on their workstation and match it back by ear. Not too much trouble if it is just a few cuts, but if the music comes from a five-minute cut and is used at every transition, then you are costing your client money. How do you think the sound guys will justify their higher bill?

When you are dubbing everything to timecoded sources, it is up to you (or your assistants, dubbers, etc., but really up to you) to make sure the quality is not degraded. Dub to higher formats—digital formats—and monitor the levels very closely. There is nothing worse than being forced to redigitize a distorted media file and finding that the dubbed source is also distorted! Some facilities make sure that everyone who dubs a source puts his or her initials, date, and machine used on every tape. Being able to track a problem back to the source does a lot to ensure quality control.

DUBBING WITH TIMECODE

There are some basic requirements for good timecode that are very important when dubbing sources to be used in an online session. The first requirement is: Never dub timecode. That seems like an outrageously stupid assertion, but one that has confused people for many years. You should not dub timecode—you should always regenerate it. Timecode is a square wave signal, and straight dubbing tends to add noise to the signal. Eventually (sooner if the timecode wasn't great to start with), the signal becomes slightly rounded and suddenly, where that sharp edge in the signal delineated a specific number, it is now just a little too rounded to read accurately. You may have been dubbing timecode for years and sending your problems downstream or, luckily and more probably, your record deck has been left in the position for Regen. This is a switch under the front Control Panel of the deck that allows you to regenerate—re-square—the signal as it passes through the deck's electronics and lays it to tape.

It may be important to keep the timecode from the original source for accounting reasons, so the stock footage can be accurately paid for or because of the way the selects reel will be made. When making a selects reel, you can be dubbing many shots from separate reels onto one reel for the convenience of digitizing or because you will lose access to the originals during post-production. The shots on the selects reel may need to be traced back one more step to the original tapes before the online. Now you should set the internal/external timecode switch on the tape deck to "ext" for external source. Make sure that the source deck and the record deck have timecode cables connected between them so that you can jam sync the timecode to the new tape. Give yourself plenty of preroll because there is a break in the timecode at every shot. You do not want the Avid or the edit controller to rewind over a break in the timecode when cueing up for a shot.

All edit controllers, whether they are part of the Avid software or part of an online suite, always expect a higher number to come later on the tape. The timecode is expected to increment continuously! If you put hour 10 before hour 5 on the tape, anyone who uses this tape will be confused. Edit controllers will search and fail to find preroll and edit points. Your name will be used in vain and no one will ever return your phone calls again.

If the beginning shots of the tape are drop-frame, then everything that is dubbed to that tape should be drop-frame, too. The Avid system associates a tape name with whether an entire tape is drop or non-drop. If you mix the types of timecode, you have to give the tape two names—very confusing and not recommended!

If you can lay down new timecode instead of taking it from the original source, you can keep the timecode nice and neat on the selects reel. If the "ext/int" switch is back in the "int" (internal) position, the record deck always generates its own timecode. Another beauty of leaving your deck in Regen is that any assemble edit (an edit that changes the timecode on the tape) picks up where the last timecode on the tape left off. During the few seconds preroll to make the assemble edit, the deck reads the timecode already on the tape. At the edit point, the deck neatly regenerates timecode without missing a frame. Either you should be purposefully presetting the timecode on the deck, jam synching the code from the source tape, or regenerating it from the timecode that is already on the new tape.

Timecode is just another tool and can be used very effectively once you know how to preset it and actively use it. If you are making a selects reel, choose a timecode that is not used by the other field tapes. If the last field tape is tape 15 and uses timecode hour 15, then start your selects reel at hour 16. No matter how a tape is named in the EDL, you can trace it back to the right tape based on the hour of the timecode.

TAPE NAMES IN EDLS

Tape name is incredibly important to the Avid and, if you get almost nothing else right, at least this should be perfect. Every tape needs a unique name. A tape must be named while digitizing or logging in the correct project. It is good practice to make the tape name match the timecode hour and never repeat a tape name. As mentioned before, but bears repeating, tape names should be short. They should be five to seven characters maximum to allow a space for a "B" to be attached to the end when making dupe reels for CMX and Grass Valley Group (GVG) systems. This chapter will discuss dupe reels later—you can't count on not needing them!

Assuming that all timecodes in the Avid refer back to real sources and all tape names are unique, let's look at the basics for preparing a list. The Media Composer has been shipping with an extra piece of software for the last few years called EDL Manager. It also ships with Avid Xpress and Symphony; if you don't have it on older versions, check the box your User Guide came in and you will probably find it there neatly shrink-wrapped. This piece of software is now part of the installation CD-ROM and is meant to be able to open bins and locate the sequences inside the bins to create EDLs.

EDL MANAGER

EDL Manager is Avid's EDL generation and list-importing software and is separate from the Avid software for several reasons. The first reason is so that it can be updated and improved upon with a development schedule separate from the editing systems. EDL Manager can then skip the long and complicated beta testing programs in favor of quality assurance just for this small piece of software. The offline editor can carry it to the online suite or the online editor can have a version running in the suite already. The offline editor can be spared some of the responsibility for making a good list if he or she brings a sequence bin to the online session with the (ideally) final version of the sequence.

With present versions of the Avid, the EDL Manager launches automatically when you choose EDL from the Output menu. If it doesn't launch, there are probably two things wrong: First, the EDL Manager application must stay at the level of the Macintosh file hierarchy, or inside the Avid folder on NT, where it was originally installed. In other words, it cannot be inside another folder hidden away somewhere. It must be on the internal Macintosh drive on the highest level, visible after you double-click on the drive icon, and inside a folder named "EDL Manager ƒ." On NT it is inside the EDL Manager folder at the same level as other Avid Utilities. You may get a message that you need to navigate to where the EDL Manager actually exists. If not, you will have to find it yourself and put it in the right folder.

Second, if it is there and still not opening, you may not have enough RAM for it. EDL Manager should work well with about 12 megabytes of RAM, but if you are working on very long sequences you have to give it more, much more. Some people give EDL Manager 40 megabytes of RAM and then never have a problem with any sequence, but they must have the extra RAM available if they are simultaneously running the editing software. The NT operating system automatically allocates RAM to an application as it is needed. Although you may still not have enough RAM, this is not something you need to set. Of course, all new systems are shipping with more RAM these days, so this problem is less likely.

If both applications are running, you can use the simple arrow icon in EDL Manager. The sequence, which is in the Record window of the Avid editing software, will load into the EDL Manager with a single click. You can also use the arrow to go the other way and turn an EDL into a sequence, but since there is more art than science to that technique, we will discuss it in Chapter 10.

If you are using earlier versions of the Media Composer, the EDL Tool is built into the Media Composer application. With EDL Tool, you are missing several important Sony list types, a standard way to get a well-tested, preread function, and a superior EDL import ability. We will talk about the importance of all of these later in this chapter, but with EDL Manager, you are working on a version of software that has been updated and beta tested for years longer than the EDL Tool. If you have a choice, use the EDL Manager.

If you are using the MCX for NT, you have the MCX EDL to use for the same purpose. This application is actually a modified version of a well-known, popular program called EDLMax written originally by Brooks Harris. Harris modified it to make it more compatible with the MCX-NT and OMFI. This discussion concentrates on features that are important to both pieces of software and to any EDL, although the actual menus are different with MCX EDL.

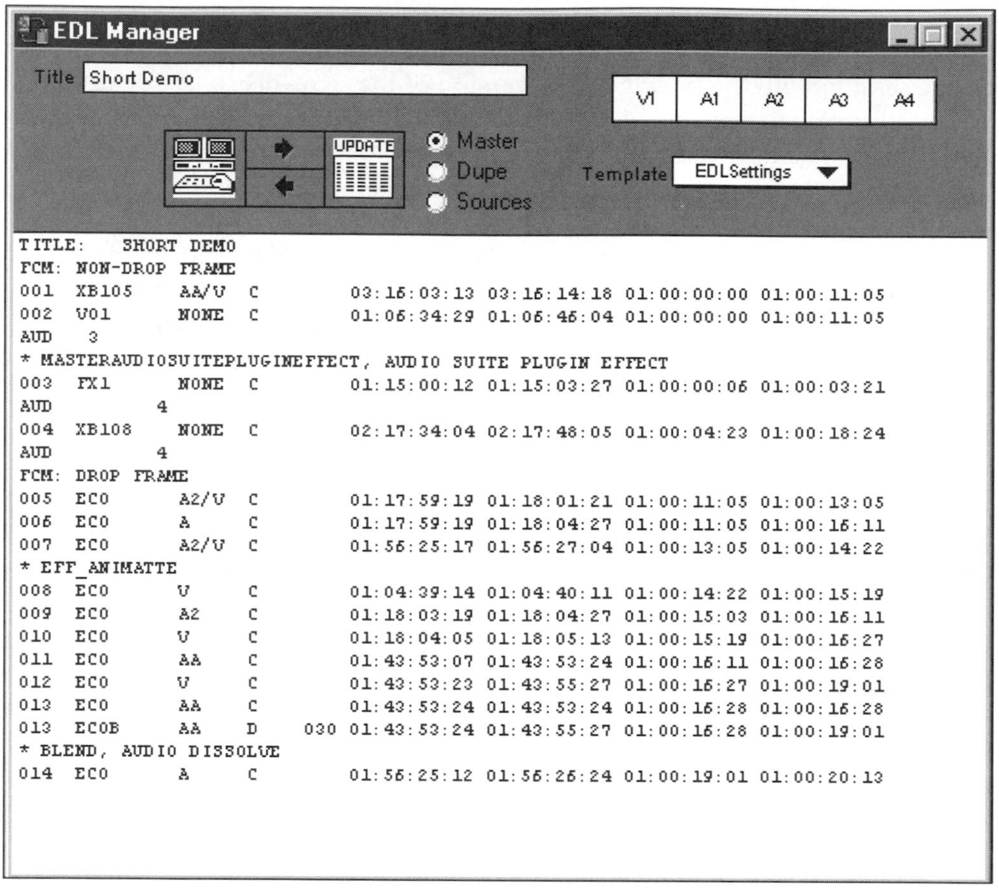

```
EDL Manager                                                    _ □ ✕

Title  Short Demo                                    V1  A1  A2  A3  A4

         ▤▤  ➡  UPDATE  ◉ Master
         ▤▤       ▦▦    ○ Dupe      Template  EDLSettings  ▼
                         ○ Sources

TITLE:    SHORT DEMO
FCM: NON-DROP FRAME
001  XB105     AA/V   C      03:15:03:13 03:15:14:18 01:00:00:00 01:00:11:05
002  V01       NONE   C      01:05:34:29 01:05:45:04 01:00:00:00 01:00:11:05
AUD   3
* MASTERAUDIOSUITEPLUGINEFFECT, AUDIO SUITE PLUGIN EFFECT
003  FX1       NONE   C      01:15:00:12 01:15:03:27 01:00:00:05 01:00:03:21
AUD       4
004  XB108     NONE   C      02:17:34:04 02:17:48:05 01:00:04:23 01:00:18:24
AUD       4
FCM: DROP FRAME
005  ECO       A2/V   C      01:17:59:19 01:18:01:21 01:00:11:05 01:00:13:05
006  ECO       A      C      01:17:59:19 01:18:04:27 01:00:11:05 01:00:15:11
007  ECO       A2/V   C      01:55:25:17 01:55:27:04 01:00:13:05 01:00:14:22
* EFF_ANIMATTE
008  ECO       V      C      01:04:39:14 01:04:40:11 01:00:14:22 01:00:15:19
009  ECO       A2     C      01:18:03:19 01:18:04:27 01:00:15:03 01:00:15:11
010  ECO       V      C      01:18:04:05 01:18:05:13 01:00:15:19 01:00:16:27
011  ECO       AA     C      01:43:53:07 01:43:53:24 01:00:15:11 01:00:16:28
012  ECO       V      C      01:43:53:23 01:43:55:27 01:00:16:27 01:00:19:01
013  ECO       AA     C      01:43:53:24 01:43:53:24 01:00:16:28 01:00:16:28
013  ECOB      AA     D   030 01:43:53:24 01:43:55:27 01:00:16:28 01:00:19:01
* BLEND, AUDIO DISSOLVE
014  ECO       A      C      01:55:25:12 01:55:26:24 01:00:19:01 01:00:20:13
```

Figure 8.1 EDL Manager

GETTING READY

There are so many variables in the linear online suite that you can never be too careful about preparing for any eventuality. There are many different kinds of edit controllers on the market today and, to some extent, they are not compatible. The main ones in use are CMX, Grass Valley Group (GVG), and Sony. There are also many other less prevalent models like Ampex ACE and Abekas; however, within your market there may be an exception to this, and you may find that the most popular edit controller is something else. In Chicago, for instance, you will find many Axial controllers.

The first rule of EDLs is this: Find out which edit controller format the linear online facility needs. Notice I said "what they need" not "what edit controller do they have," which is for a very important reason. They may have a CMX

Omni not a CMX 3600, but the lists are mostly compatible, so the facility may ask for a CMX 3600 list instead. Or they may have a bizarre 10-year-old Brand X model that went out of business seven years ago, but the darn machine just refuses to die. They may ask for a CMX 3600 as well because out of necessity they may have or may have written their own software to translate from CMX 3600. Most people consider the CMX 3600 the Latin root of all list formats.

FORMATTING FLOPPIES AND RT-11

You must format the floppy disks for CMX and Grass Valley Group edit controllers in an unusual way. CMX originally used a disk format that was useful with the DEC computers they used internally when they were developed. Because GVG developers came from CMX, Grass Valley Group also uses a version of that original CMX format. The format is called RT-11 and is problematic with both Macintosh and Windows because these operating systems are not designed to recognize it. Special commands must be written to make the computer ignore the fact that it can't read the RT-11 floppy disk you just popped in. The edit list software must take over for both reading and formatting these floppies. This is why if you just pop a CMX disk into the Macintosh, you get a scary message saying this disk is unrecognizable and do you want to format it? No! Eject, now! The Mac defaults to this evil question whenever it cannot recognize the formatting on a disk. You can only work with RT-11 disks when you are in the Avid, EDL Manager, or another piece of specialized software designed specifically for EDLs.

RT-11 was not written to make your life a living hell, but it sure doesn't help. Add to this the restriction that the disk used must be a double-density, double-sided disk. I want to say that again: double-density, double-sided. Look at all of the floppies strewn around your computer. I'll bet none of them used for standard, everyday saving is double-density, double-sided. They probably all say HD, which stands for high-density. The disks you want should say DD on them. They are not interchangeable! This is an example of dedicated hardware being passed by very quickly by technological standards. Double-density disks hold only 800 KB of material. This could be a small novel, but it is not very many EDLs. The real world uses HD floppies because they hold 1.4 MB, almost twice the capacity of the DD disks. Even the HD disks are pretty pathetic in a world where gigabytes are becoming the standard measurement for storage capacity. If you are making the edit list, it is up to you to scrutinize every disk that is handed to you.

I strongly recommend that the DD disk should be unformatted or already formatted for RT-11 from the edit controller that will use it. The formatting of the disk for the EDL has caused many people the vast majority of their EDL woes.

Some early versions of EDL Manager did not let you format RT-11 if the disk was formatted for anything else already. That was changed for version 1.9.1v2 and later. No matter which version of EDL Manager you may have, there is a trick for unformatting a floppy. First, quit all the programs you are running. Pull down the Erase Disk command under the Macintosh Special menu, as you would normally erase all the material on the floppy. Let the process run for a few seconds, but before it finishes you must force the Macintosh to crash. The only way you may be able to interrupt the erase is to press the Restart button on the outside of the computer case or press the Control-Command-Power keys all at the same time. This works to totally screw up the format on the floppy, but it sure is ugly. For all of the NT versions of EDL Manager and later versions on Macintosh, this method is not necessary since the system will competently erase any original formatting and replace it with RT-11.

So now, you have an unformatted double-density, double-sided floppy. If you pop it back into the Macintosh, the system tells you the floppy is now unrecognizable—this is one of the few times you want to see this message! The NT will give you a similar message. Now the floppy can be reformatted while you are running the EDL Manager or other EDL software. Is it any wonder that busy post-production facilities would rather FedEx you a few of their own disks? Overnight mail with a proper set of disks is a good idea from their point of view, so give them the chance by letting them know you are coming far enough in advance. Sometimes there are differences that have developed over time between the disk drives at their facility and your brand-spanking new Macintosh G4 floppy drive. Old drives get cranky and may occasionally not accept a floppy that has been formatted by another drive. The busy facility has also seen more than their share of HD disks with EDLs and offline editors who blame the Avid! If all else fails, to get your floppy to be accepted by their online edit controller, copy it to another computer that can read RT-11. This other computer may potentially have EDL Manager running. Turn the EDL into a text file using a word processing program, and copy it to a properly formatted disk.

SONY AND DOS

One way to avoid all of this is to work with a Sony list. Sony saves to a DOS-formatted disk that the Macintosh can easily handle. Early versions of the Macintosh needed one of two different extensions, either PC Exchange or DOS Mounter. These extensions need to be enabled in the Extensions folder to read or format a floppy for Windows. You would choose "Other" during the formatting dialog on the Macintosh and choose the proper format for the size floppy you have. Early Sony edit controllers may read only DD disks, but the later ones read HD. If in doubt, get the DD and format 800K.

REAL-LIFE VARIATIONS

For one final twist, there may be weird combinations of formats for disks and formats for edit controllers. The later CMX controllers, like the OMNI, can accept DOS-formatted disks. The production company may have its own list optimizing software that runs on a DOS or Windows system, so they may want the list, no matter what it is, on a DOS floppy. Some older edit controllers still work from 8-inch floppies. Most people have never even seen these before. Forget about compatibility. These unlucky people have devised a way to send information serially, through a cable from another computer that can handle the 3.5-inch floppies to their online system that can only handle 8-inch floppies. EDL Manager allows this, too, but it takes a 1-2-3 Go! procedure that you probably don't want anyone to see.

BE PREPARED

So there appear to be many ways to get this whole EDL process wrong, because there are so many ways to get it right. With all of this potential for 9 A.M. thrash at the online facility, there are ways to cover yourself. First and foremost, you should ask the production facility what they want. Asking far enough ahead of time gives the facility time to send you their pre-formatted floppies just in case. If you are using the same production houses repeatedly, then you know their drill and can have procedures in place for each edit suite of each production company you use.

But don't bet everything on one floppy! Make two, they're cheap! You may also want to save the EDL in several other formats. Always make exactly what the facility asks for and then make a CMX 3600 list. Pick your market's favorite format and make that your backup format. It is up to you whether you want to put it on DOS or RT-11. (If you choose both, it could make you look pretty slick.) Then you should save the sequence bin to a Macintosh-formatted HD floppy and, if the project was small enough, include all the bins. You may find yourself actually making the list again from the sequence at the online facility's Avid or from your own copy of EDL Manager (which you brought with you). After all of this is done, trust nothing! Make two printouts and bring a copy of the project on VHS or some other format that can be played back in the online suite without a charge for another deck. You or your producer should control the playback of this dub in the suite so you can follow along with the online assembly.

WHAT IS AN EDL?

Now that you are prepared to make the EDL, let's take a look at what an EDL really is and why it is so hard to make one from a nonlinear editing system. The only information the edit controller wants from the Avid is the tape name and timecode (sound familiar?). The edit controller cues up the source tape and inserts the shot onto a specific timecode on the master tape. So a basic event — a cut, for instance — needs four timecode numbers: an inpoint and an outpoint on the source tape, and corresponding in and outs on the master. It looks like this:

```
017  100   V   C   02:00:24:19  02:00:25:14   01:00:22:02  01:00:22:27
```

When you need to make a dissolve with analog tape, most of the time you need a second source. If you must dub a shot onto a second tape, it is typically called a B roll. Thus, an A/B roll system is one that allows dissolves. A dissolve needs the inpoints and outpoints from the B roll in the event and looks like this:

```
012   100   V C        02:18:31:29  02:18:31:29   01:00:12:19   01:00:12:19
012   100B V B   030   02:23:26:15  02:23:28:26   01:00:12:19   01:00:15:00
BLEND_DISSOLVE
```

Any kind of fade up from black or from a graphic that has no timecode is an event that starts with black (BLK) or an auxiliary source (AX or AUX). It looks like this:

```
002    AX    V   C        00:00:03:27  00:00:03:27  01:00:01:28  01:00:01:28
002    100   V   D   030  01:43:44:28  01:43:50:29  01:00:01:28  01:00:07:29
BLEND_DISSOLVE
```

So all of your edits that come from a non-timecoded source, like CD audio or a PICT file, come in as AUX. If you have 100 PICT files, this could be a problem.

TRANSLATING EFFECTS

Any time you create an effect like a wipe or a 3-D warp, it must somehow be translated to this very limited, older format that is concerned with tapes and timecodes. There are standard SMPTE (Society of Motion Picture and Television Engineers) wipe codes, but some of the most popular effects switchers do not use them. EDL Manager has a list of the most popular switchers represented, and you can use these simple templates to represent your wipe effects. You can even take these templates for the switchers and modify them to your own switcher if you are feeling geeky. The wipe codes can usually be found in a manufacturer's technical manual, and these templates are just text files.

There is no way to create a generic 3-D effect that any group of DVEs (digital video effects generators) made by different manufacturers can read. Many people don't have the capability to send DVE information from their edit controller to the DVE. These two machines could easily be from different manufacturers and have no communication between them except "Go Now." Different DVEs use different scales for computing 3-D space, their page turns look incredibly different (if they can even do a page turn), and really, forget it, there is no standard for effects. There will not be such a standard for a long time to come. You have the timecodes for the two sources for the effect and you have the VHS. If you are really persnickety, you may have even written down some of the effect information as you made the effect. Adding a comment to the list like, "second keyframe is exactly 20 frames later" doesn't hurt.

This information is best delivered to the EDL itself with the Add Comment feature in the Avid. If you highlight a single shot in the sequence with the segment arrow, you can pull down the menu above the Record window and request the Add Comment option. The comment shows up in the EDL as a little bit of extra text when you choose Show Comments in EDL Manager. You can search by comment in the sequence using Command-F. Think of it as locator information that makes it to the EDL. Now you can enter information about the sequence or the transitions.

Even if you can represent exactly the way an effect was going to happen, you cannot foresee how the online edit suite will be wired. The effect may call for three levels of dissolves, but the switcher only has two levels it can do simultaneously. Your effect may require that the editor create multiple layers of video, which the editors should know how to do quickly and efficiently with their own configuration of equipment. If you try to be too specific about the exact procedure for creating an effect, you are denying the online editors the chance to do what they do best.

MULTIPLE LAYERS OF GRAPHICS AND VIDEO

The final insult in translating from nonlinear to linear is the way most EDLs represent multiple layers of video. They really don't. This means you need to reduce all multi-layered sequences to a series of single video track EDLs. The nested tracks must be broken out of track one and put into separate EDLs for each track in the nest. This provides you with the ability to run more than two video sources simultaneously in the online edit suite. Don't count on it unless you have researched the specific suite you will be using.

The online editor loads the video tracks as separate sequences into the edit controller and combines them. You can help the online editor figure out which tracks go where by the way the separate lists are numbered. Each edit is called an *event* and has a number. If the first layer of events ends at event 110, then all of the events for the second layer of video can start with event 200. The third layer can start at event 300, the fourth layer at event 400, and so on. The editor knows that when two events overlap in the EDL and if one has event number 50 and the other is event 210, then 210 is part of a multi-layered effect.

Any multi-layered sequence should be carefully discussed with the individual online editor. Make sure they can even get close to what you are doing on the Avid. You may find they say, "Sure, we can do that—in the Flame suite!" Listen for the cash register sound when they want to put you into a special compositing suite. There may be no other choice but to go this route. On the other hand, a slight redesign and some extra goofing around may save you thousands. A freelancer using Adobe After Effects or another desktop compositing program may be the right answer if you have the time to render. There is a tradeoff between waiting for a short render in the expensive compositing suite and waiting until Monday for After Effects. With the use of render acceleration boards like ICE (Integrated Computing Engines) for Adobe After Effects, the render times on the desktop are becoming more competitive. It may make sense for you to get a cheaper online suite and arrive with all the effects already made so you just slug them in. Did I mention that you can't make changes in the online suite?

SIMPLIFY THE EDL

If you try to make an EDL from a complex sequence with layers and nesting and you just load it all into the EDL Manager and press the purée button (or Update), what comes out will not be very useful. You may get some error message as the system warns you that what you are trying to do is not very nice. This is usually some kind of a parsing error or a message that an area of the sequence is too complex to be represented. Make a copy of your sequence and start to simplify.

First, get rid of the nesting. Figure out what are the most important sources for that effect. What timecodes do you need? Keep just that source. If you absolutely have to have the timecodes from multiple sources for that effect, you should not have nested it in the first place. Yeah, right, that assumes you knew you were going to an EDL. If you are going to an EDL, do not nest. Chapter 10 describes how to unnest. You enter the extra timecode and source information by hand as a Comment. In EDL Manager, you choose which video track to use, starting with V1.

If you know you are going to an EDL, don't use imported graphics. An online suite may be able to use PICT or TIFF files through a sophisticated character generator, but if you know you're going to an online room, all graphics should be fed to you on tape. The timecode numbers of the fill, matte, or backgrounds can go cleanly into your EDL with few questions asked. Also, if you use imported graphics, it means additional work for the graphics department because they must output the electronic version for you, then they have to make a fill and matte and lay it to tape for the online. If you have subtle variations in logos or graphics, the online editors can put in the wrong graphic since now you expect them to figure out which graphic on the graphics tape corresponds to your non-timecoded source. Even if you use imported graphics to simplify your edit or because you didn't know you were going to an EDL, I strongly recommend recutting those sections with the same graphics tape that is used for the online.

SOUND LEVELS

How did you change your sound levels when you were mixing? If you used rubberbanding, there is no problem. Those levels are not in the EDL, but the sound edits do not confuse the online editor either. If you used the older method of creating Add Edits, changing the levels, and adding dissolves to smooth out the level change, then you have some work to do. These Add Edits are real edits and the EDL Manager is smart enough to ignore them, unless you have added a dissolve. The EDL Manager puts an edit in the EDL and tells the online editor to dissolve to a B roll, a dub of the exact same audio source, just to change the audio level. This is confusing and annoying. You need to remove the dissolves to any audio transition that is just for a level change. This is actually easy because you can go into the effect mode and lasso or Shift-click them and delete them all at once. This is a good reason to use rubberbanding. You can also use a new feature in later EDL Manager versions called "Audio dissolves as cuts."

For the high-end jobs, the sound is usually sent out for audio sweetening and is laid into the online session in one big edit. For lower cost jobs, a digital cut is laid to tape (sometimes split tracks, sometimes stereo split) along with the low-resolution video. Then the online editor cuts his or her video on top of that, covering it. For these EDLs, you should disable the audio tracks, creating a nice, clean video-only EDL. If you want to be safe, you can make an extra EDL with audio and video. This is just in case they need to find the video for a sound bite that has been entirely covered up. It will also help if there is a problem with the sound from the Avid in a certain spot.

SETTINGS

There still seem to be many choices in the EDL Manager that haven't been described yet, so don't get nervous. Most of these functions are for streamlining an auto-assemble and other more obscure situations. We will get to them at the end of the chapter, but for now let's go with the simple answers.

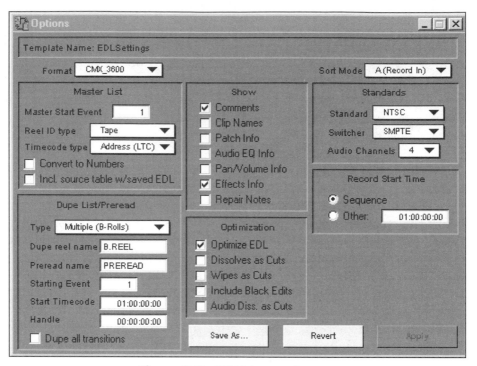

Figure 8.2 EDL Manager Options

SORT MODES

A sort mode is a way of ordering the events in an EDL so that they can be edited out of linear order. Linear order means that you load a tape into the deck, adjust the levels from the color bars at the beginning of the tape, fast forward to the spot on the tape where the shot is, and perform the edit. Then you load the tape for the next shot in the program and do the same thing. You do this whether the source tape is 20 minutes long or 2 hours long. There are a lot of repeated actions and wasted time shuttling the source tapes back and forth. Unfortunately, a linear order is the easiest to follow and the safest way to assemble a show and be able to compensate for a mistake or make changes. You know immediately if a shot is missing or a sound bite is wrong because it does not make sense as you put it in place.

The time savings are significant if you can assemble a show out of linear order. It means you only need to load a source tape once, do all that color bars setup, and skip through the show, dropping shots wherever they are needed from that source reel. Then put the tape away. You might have to bring it back for effects later, but if your show is just cuts and dissolves, you have made serious savings in time and money. But imagine making a change that makes the show just a few seconds longer. If you have been inserting shots all over the master tape out of order, this is pretty serious. How can you ripple that change and push down all the edits after it in the program if there are already 100 shots laid down after the change? You can't really; if you are working with digital tape, you can clone the master and make an exact digital copy. If it is an hour-long program and you are halfway done, then kiss goodbye a good 45 minutes of set-up and dubbing time. Did I mention that you can't make changes in online?

This kind of penalty for being wrong or making changes at the last minute tends to frighten even very brave online editors. For this reason, they want to be the person who makes the decision how to sort the list. When asked, they inevitably tell you to use the A sort mode because that is the linear mode and they sort it themselves in their edit controller to make it a B, C, D, E, or S mode.

Let's do a quick recap of what the different modes are so you can discuss them intelligently at cocktail parties. B mode is for long-format shows where the source tapes are shorter than the master tape. The source tape fast forwards and rewinds and finds shots that go in a linear order on the master. This is easier to follow because the master moves forward while a specific source tape is being used. As soon as a new tape is loaded, the master rewinds to the first place the new tape is used. It then starts to move forward again until all the shots from that source are inserted on the master.

A C-mode list is for when the source tapes are longer than the master tape. You may have a 90-minute film transfer for a 30-second commercial. (I hesitate to call them "spots" anymore since in the UK that means pimples!) Here, the source tape always moves forward in a linear way and the master tape flails around rewinding and fast forwarding to assemble out of linear order. This is potentially faster than B mode but harder to figure out what the heck is going on. Very disconcerting for clients who should just look away if they are feeling queasy.

D and E modes are like B and C modes except that all the dissolves are saved for the end. This works only if you have time base correctors (TBCs) that have memory for individual tape settings. More common these days is an SDI (serial digital interface) signal from a digital deck. Supposedly, the SDI signal does not change from deck to deck like an analog signal and so, theoretically, never needs adjusting. The editor must match a frame perfectly from something that he or she adjusted this morning or yesterday so that you don't see where the pickup point is to start the dissolve. D and E modes get used more in PAL countries where they don't worry about analog hue adjustment, unless it's terribly wrong to start with!

S mode is useful for getting a list that will be used for a retransfer from film. If you have been cutting with a one light or a best light, inexpensive film transfer without shot-by-shot color correction, then you want to go back and retransfer the shots that were actually used. S mode is the order that the shots came on the original film reels and allows the colorist to quickly identify only the used shots for the final color correction.

DUPE REELS

After choosing the format and the sort mode, it is time to think about dupe reels. What are you going to do when you need to dissolve or wipe? If you haven't experienced the difficulty of dissolving between two shots on the same tape, then you are about to. With a traditional analog tape suite, you must physically copy the second shot onto a dupe reel so that you can roll two tapes at the same time. Seems kind of old-fashioned and quaint now, but a dupe reel is still the necessary evil in many suites across the world. The biggest change to this procedure was *preread.* Someone figured out that since the digital format tape was just outputting digital information, short amounts could easily be held in a buffer, or temporary memory, in the record tape deck. During the preroll, the master record machine loads the digital images of the shot already on the master tape into the memory and plays it back as a source deck. With the last shot held in memory, the master tape switches from being a source and records the dissolve with the new, incoming shot.

Some editors also use a kind of non-destructive preread called *auto-caching.* Auto-caching holds the end frames of a shot for dissolves and various layers of multi-layered work in a cache that uses a digital disk recorder (DDR) or a digital tape. The edit controller figures out what needs to be cached for later use in a dissolve or effect. Then it automatically lays the image on the DDR or digital tape. When the tape with the other half of the dissolve is put into a deck, the controller performs the dissolve between the tape and the auto-cached piece of video. This

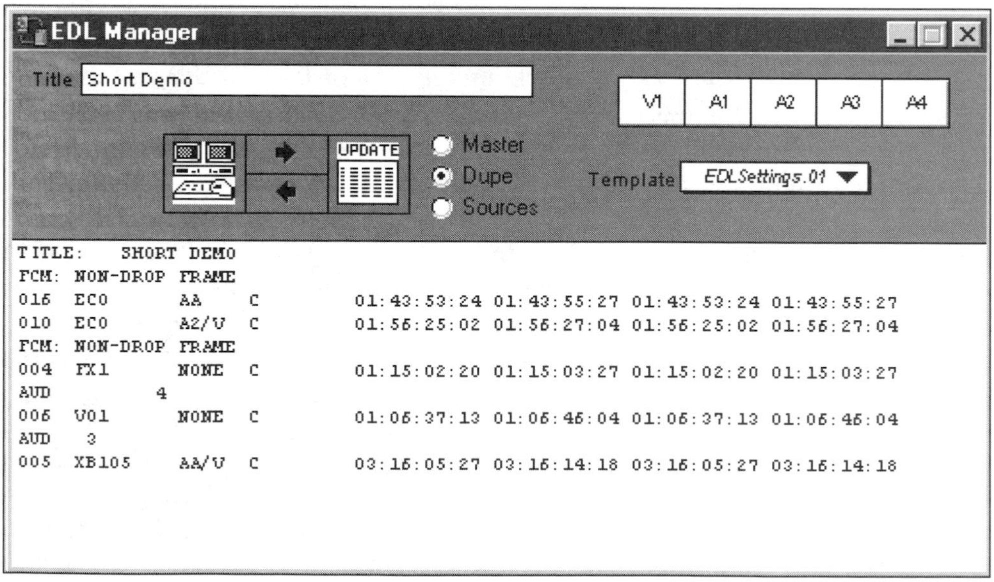

Figure 8.3 Dupe Reel List

works better than preread because prereads are destructive, meaning you are committed to that edit permanently because the last image is recorded over itself when the recorder becomes a source. You only get one chance to do a preread correctly, and then you must re-edit the previous shot. Auto-caching uses a preread EDL to create the auto-caches that can be executed and re-executed if needed.

Most of the choices that deal with how a tape is dissolved are refinements on these basic requirements for either a dupe reel or preread. Under certain conditions, you can significantly reduce the amount of tape loading and shuttling, but most of the time you choose the simplest technique. The online editor has a very specific requirement and again, as with sort modes, can modify the simplest format into something her or she can use best.

The simplest, best answer for most circumstances is the Multiple B Rolls option. It is also the least clever, but with more sophisticated editing systems out there, editors can modify this list to speed up the assembly. The Multiple B Rolls option assumes that there is a separate B roll or dub of every tape in the show. This is probably not true unless you have a very small amount of material or have made simultaneous duplicates during a film transfer. It is more likely that the online editor looks at the list, sees that a dub is required, and hunts down the original tape. The timecodes on the B roll are the same as on the original tape, and the name of the B roll is a variation of the original tape name. Finding the shot is easy because it has a "B" after the tape name or the tape number is incremented by 500, depending on the type of list format. The editor can then just dub that shot when it is needed or change it over to preread in the edit controller.

Making a Special Dupe Reel

The other B roll choices allow the editor to prepare before the online session by taking all the shots that need to be dubbed and putting them onto one reel. This keeps the tape loading down to a minimum and allows you to move faster with fewer decks configured (and charged for)

The most dangerous dupe reel choice of all is the One, New Timecodes option, although it is extremely popular in Europe. This requires that you make two lists. The first one is for the master tape, and the other is for the tape where all the B shots will be dubbed. This new dupe reel has perfect, new timecode because you are ignoring the original timecodes of the source tapes and starting all over again. You take each tape and dub one after the other onto this new reel, which starts with timecode that begins at 1:00:00:00 (although you can change this). This means that the timecode link between the original source tape and the dupe reel will be completely broken as soon as the dupe reel is made. The dupe reel name, which defaults to B.REEL, shows up in the list every time a shot needs to be dubbed for an effect. You cannot look at the EDL dupe reel name and figure out where the shots came from originally. Many people choose One, New

Timecodes and then do not make the dupe reel list. This is a big mistake. When the online editor turns around and asks, "OK, where is this B.REEL?" you really should have another list to hand over (and maybe even the premade reel). Unless you make the B.REEL yourself, that reel needs to be assembled at the beginning of the session.

Because it is assembled before the editing really starts, changes to the B.REEL are also very difficult. You *must* make a second list for the dupe reel! If you must make a dupe reel, then One, Same Timecode is better. If you forget to make the list for the new dupe reel, at least the original timecodes are in the list and, if the timecode hour matches the tape name, finding the original is a snap.

Another dangerous choice for the unsure is None. This gives you a list that most edit controllers cannot read, which at first seems like a rather bad choice. Not so if you are going to another NLE or an effects box like the Discreet Logic Flame or Quantel Henry. These systems do not need to dub a tape in order to dissolve because, once digitized, individual shots are divorced from the physical restrictions of tape. Loading the list into the NLE systems forces it to see the tape 001 and another tape named 001B. Is this the same tape? There is no way to know for sure, so the NLE system makes the safe but dumb assumption that 001 and 001B are two distinct tapes. When you are digitizing this material, the system asks for tape 001 and then proceeds through this tape to the end and grabs all the shots that are needed for the sequence. Then the NLE system asks for tape 001B, rewinds, and does the same thing again with the same tape! This is not efficient and, at those room rates, you need to be efficient. The None choice shows just one tape, 001, and the NLE system requests that tape only once.

PRINTING THE LIST

When you print out the list, you should make two versions. Once you are ready to print the EDL, think about who will be using the printout. The people supervising the online edit certainly need it in front of them. They may not be the most list-savvy people in the production, so there are ways to make the list more readable. Include comments in the list when you are about to print. Use comments like "clip name" and the source table, which is a list of all the sources needed for the online assembly. Then print a stripped-down version for the online editor. Again, ask if there is any special information the editor wants, like audio patch information or repair notes. I know some offline editors are reluctant to have repair notes show up in their list because it looks like they did something wrong. Repair notes are important because it may just be a case of the effect being too complex and not being able to represent it correctly. If you have any repair notes, then the online editor should see them.

THE SOURCE TABLE

The important function of the source table is when occasionally the EDL Manager finds a tape with a name that is too long and must shorten it. Sometimes there may be a duplicate tape name from another project. There may also be two separate versions of the same project that have been combined. The EDL Manager has to change that tape name in the list. The source table tells you that since you had a tape 001 from project X and a tape 001 from project Y, that, with a Sony format, EDL Manager changed the second tape name to 999. A CMX list changes a second tape 001 to 253. According to the system, these are two different tapes! This is why the original logging is so important. If you don't have that information, then you could get very lost. Never attach a source table to the electronic version EDL on the disk because it is really just a very long comment added to the last event. Many systems have a limit as to how many comments they can read per event, and this will cause an error in reading the list if you have a lot of tape sources. Make the electronic version without the source table, but always print it out separately to take with you.

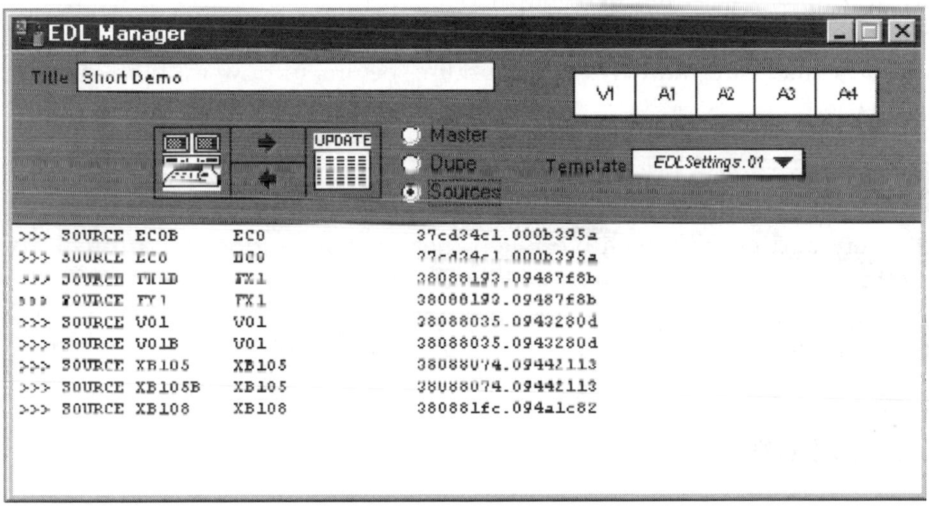

Figure 8.4 Source Table

PAL LISTS

If you have a PAL project, be aware that the very first time you open EDL Manager, it defaults to NTSC. You can change this to PAL under the Standards Setting and it will stay that way. PAL lists become complicated when you are working in film and want to go back to the negative because of the two different methods of dealing with the difference between 24 fps and 25 fps. Sometimes you slow down the audio and sometimes you include extra frames in the video (that pesky 25th frame). The PAL Method 1 is for film projects in which video and audio are transferred together at 25 fps. PAL Method 2 is for when video and audio are digitized separately and synched in the Avid. If you have a video project without any matching back to film, then choose plain vanilla video PAL.

MATCHBACK

Matchback is a method preferred by low-budget films or film projects that are primarily for video distribution but will be projected sometime in the distant future. This is an option for both Media Composer and Avid Xpress that will output a 30 or 25 fps project as a 24 fps cut list. Film transfer to video adds extra frames. There is no way for the video-based project to keep track of the A frame, so every outpoint in a matchback project can only be accurate to plus or minus a video frame in NTSC. The matchback is much more accurate in PAL. For many people, depending on the material, this is quite acceptable compared to the cost of conforming a negative first and then transferring to video or using a Film Composer. They get a video project at a high, two-field AVR that they can shop around or distribute, and when the funding begins to flow in, they can get the negative cut.

Using Symphony Universal at 24P allows the creation of both a high-quality output when working at 24 fps and a perfect negative cut list using FilmScribe, a cut list utility. This will allow you to broadcast the finished project today and cut the negative when theater distribution is required. Using the Universal Editing option on Media Composer, you can output at 14:1 progressive (which looks pretty darn good for offline) and output the perfect list for cutting negative.

EDL TEMPLATES

It should be clear by now that creating EDLs is rather complicated. It is also clear to the folks that receive these EDLs from Avid systems that many people get it wrong. I am still amazed at the number of online facilities that do not own a copy of EDL Manager to make their own lists. The amount of 9 A.M. thrash that is completely avoided by asking people to bring only a copy of their sequence bin more than pays for the software. Of course, the translation to EDLs becomes less desirable every year as Avid solutions to Total Conform are used by more facilities.

Regardless of whether you are an offline boutique that sends the lists to another facility or you are the facility and must make different EDL versions for each of your edit suites, you could use a little help. To make sure the lists are always correct and to make sure that you can turn this chore over to even the least experienced assistant, you can make EDL templates. In the latest versions of EDL Manager (version 7.1 and later), there is the ability to take any setup and save it as a template. Therefore, you can get the information correct once and save it for reuse. These templates are simple text files, so they can be easily copied to a floppy disk and loaded onto any system with the proper version of EDL Manager. As a facility, instead of faxing a complicated series of instructions, you can e-mail or overnight mail the small file on a floppy disk, which will guarantee that the offline editor gets it right.

You will also find that you may need to make multiple versions of every EDL for different reasons. You may need an audio-only EDL and an EDL for each of the video tracks used in the offline sequence. If you create a template for each of these often-used versions, you can make them all much faster. Just choose the Template pull-down menu for Video Suite 1 Audio Only or GVG Suite Video 2 and quickly save the new version.

CONCLUSION

Here is a checklist of steps that should be taken before the linear online assembly:

- Simplify a copy of the sequence before you make the list. Do this based on feedback from the online editor.
- Format the floppy for the needs of the production house (either DOS or RT-11 or both).
- Use the EDL template sent by the post facility or their faxed instructions for formats and options.
- Make any extra versions that might be important (audio only, video only, V1 separately, etc.).
- Use the same template to make a CMX 3600 version.
- Print the list twice. Make one for the editor and one for the producer. Include the source table in the printed versions only.
- Make a VHS copy of the sequence that you can control yourself in the suite.
- Don't forget to bring the source tapes!

9

Overview for Finishing

One of the most exciting aspects of the digital nonlinear technology is the promise of finishing a project completely digitally. The goal of never going to tape is a reality today for many producers, especially for newsgathering. The post-production equivalent is the ability to input everything into one system and free yourself from having to go to any other facility to output a high-quality, professional finished product. As we discussed earlier, this means mastering new skills and taking on the responsibilities that other specialists have taken care of in the past. Specifically, it means learning to set your own video and audio levels. It also means keeping the entire workflow in mind at even the earliest stages and scheduling enough time for all the details at the end of the project.

First, let's look at the specific workflow involved in finishing on the Avid or Total Conform. If you have been digitizing at high resolution throughout the project, you can output to tape and be done with it. All your levels should get close scrutiny at this stage because you are the last person to run the project through a quality check. This is a serious responsibility, and you should take nothing for granted. Symphony 2.0 gives you a breakthrough method to limit all your levels using SafeColor, which we will cover in depth in this chapter.

If you have been working at low resolution throughout the project, then there are quite a few more steps. In the simplest project, you redigitize the shots in the sequence, add in the high-resolution graphics, render the effects, and digital cut to tape. If your sequence is that easy and straightforward, this process will be simple and efficient. If you have a long, complicated sequence or some not-so-simple demands from the client, then you need the extra power and flexibility that the techniques described in this chapter give you. Some of them are basic and intuitive, and some are pretty complex. I invented some of the more complex techniques in response to real-world complications. These techniques have been previously published on the Internet, given out at seminars, and faxed by Customer Support. One of the benefits of Total Conform is that simple is good, but flexible is better.

Finishing is one situation where Media Composer and Symphony have an advantage over Avid Xpress. Avid Xpress is missing one important feature to speed the offline to online process: decompose. Although this sounds like it would smell bad, think of it as a natural step in the life cycle of the project. It gets you across the detailed transition from offline to online. You can, of course, go from low resolution to high resolution without decompose, but it is not as fast, flexible, or efficient. We will describe this feature in its proper place in the overall process.

THE MERIDIEN VIDEO SUBSYSTEM

Starting from the ground up, Avid created its own video board to edit with uncompressed YUV digital video. Avid also made it much easier to get signals in and out of the system by using most of the industry standard input and output signal types and putting them on a separate I/O subsystem. You can now input and output composite, s-video, component, and serial digital all from the same rack-mounted box. Audio can be tied to a two-channel audio card (Xpress) or the standard 8:8:8 channels (input, monitor, and output) we have come to expect from the Media Composer 9000.

Because all of this hardware is in a separate box, it is away from the CPU and thus better shielded from any electronic noise that might be generated by the computer's internal systems. It also means that a single connector to the host computer is all that is required for audio and video digitizing and video compression. This saves valuable slots and allows you to continue to add network cards, Fibre Channel, SCSI, and DVE with perhaps a slot left over depending on CPU choice. The Macintosh systems must still use a PCI Extender since the high-end workstations only contain three slots. Important Avid boards are attached through this customized, rack-mounted unit.

Now the resolutions are named by their compression level, such as 1:1 for uncompressed and 2:1 for a 50 percent compression. There is a low-quality two-field resolution called 20:1 that is mixable with the 3:1 and 2:1 and will quickly make you forget AVR 12. A 10:1 compression was added after Symphony 1.0 shipped. The single-field resolutions are marked with an "s"—as in 40:1s—to make the difference clear. As with the ABVB, Avid's previous video board, there is mixing of single-field with single-field, two-field with two-field, but no mixing at all with uncompressed. There are multiple single-field and two-field resolutions but, as of the first version of Symphony, there is no real-time play multicam (software multicam is possible) . This is planned for a later release in the near future.

Editing at true 24 frames per second has just been added with the last releases. The older Macintosh-based systems are still valuable to keep your offline costs low and then move the project to the Symphony for redigitizing and conforming. Of course, if AVR 77 is good enough for your projects now, then you can stay on the same system for offline and online. Having Meridien boards on offline and online systems will allow you to more easily share media with Fibre Channel media-sharing networks like Unity. If you need uncompressed quality or the higher quality of the new Meridien compressed resolutions, then you should be aware of probably the most important new feature for Symphony: Total Conform. We'll go into detail about that later in the chapter.

NEW COMPRESSION SCHEMES

Two very interesting aspects to this new hardware don't grab the kind of attention they deserve. The first is that the compressed video resolutions are stunning. It will take a very sharp eye pressed up against the screen to tell the difference between the 2:1 compression and the uncompressed resolution, especially coming from a very clean signal like SDI. The new Meridien video subsystem has better filtering, oversampling, and decoding, so even noisy signals compress better. Noise in the signal tends to take up valuable bandwidth with compressed images, but if you are working with a film transfer to Digibeta and can digitize using the SDI input, you may find yourself using uncompressed only for your most finicky customers.

Why then is uncompressed such a holy grail? And why do clients insist that they cannot accept a compressed signal—only Digital Betacam? Well, first, uncompressed will never introduce compression artifacts under any circumstances, so you will never have that rare but awkward moment when suddenly a complex scene doesn't look as good as it should. But, realistically, uncompressed images are required for many network deliverables as a format for archiving. No one wants to save a compressed version for future release or syndication because of all the upconverting that will occur when the world slowly shifts over to HDTV and DTV.

The other concern with compressed images is something called *cascading*— the further compression of the signal for transmission purposes. Broadcasters are concerned that if they compress something 2:1 and then compress it again 12:1 for satellite transmission, it will not look as good as an uncompressed signal compressed 12:1. In my humble opinion, it would seem that the best answer would be to not compress it 12:1 if they really want a high-quality image, but the economics of satellite time put the burden of quality on the production company not the network. Findings show that after the first compression of a signal, further compression cascading has little effect (Cornog, Katie. "Factors in Preserving Video Quality in Post-Production when Cascading Compressed Video Systems." *SMPTE Journal*, Vol. 106, Number 2, 1997).

Archival concerns are legitimate, so there is definitely an important market for uncompressed video; however, if you want your production to be completely uncompressed, do not use Digital Betacam. Digital Betacam uses a high-quality compression scheme that comes out at a little over 2:1 with color sampling at 4:2:2. When a client refuses to accept a compressed image and must have a signal in 4:4:4, but will accept Digital Betacam, it shows that there is still a lot of misunderstanding and ignorance about the value and dangers of using compressed video.

PICKING THE RIGHT RESOLUTION

Look at the total amount of material that you can realistically put onto the storage available to your system. At this point in the technology the limitations have much to do with the throughput of the computer buses and whether you are using SCSI or Fibre Channel. With SCSI you need to take into account your type of SCSI accelerator card (single-ended or differential) and the length of SCSI cable allowed. The rules are changing pretty fast and, with the widespread acceptance of Fibre Channel, you will see some pretty amazing drive farms. So for offline, use the lowest resolution you can stand without squinting; think of the low resolution as a new form of impressionism! If you are confident of the quality of the footage for example, a film transfer from an excellent director of photography — all you really need to see is boom shadows, lip synch, and distracting backgrounds; however, the client may determine the quality and leave you no choice but to figure out how to get all your media on your drive at that (higher) resolution. Some people, especially those not used to working with compressed images, get very nervous when viewing low resolutions and just don't want to put up with the lower quality. They must have the better-looking image throughout the project.

With the NuVista+ video card you cannot mix AVRs at all. Once you choose an AVR, stick with it throughout the offline or online part of the project. With the Avid Broadcast Video Board (ABVB), you can mix single-field AVRs (2s, 3s, 4s, 6s, 8s, 9s) with other single-field AVRs, multicam resolutions (2m, 3m, 4m, 6m) with other multicam resolutions, and two-field AVRs (12, 70, 71, 75, 77) with other two-field AVRs. This means that with the ABVB, you can digitize all your timecoded footage at AVR 3s and import all your graphics at AVR 9s. Many people like the option of AVR 12, which is a low-resolution, two-field AVR. Using AVR 12 at the offline stage allows you to digitize all the timecoded footage at low resolution but import all the graphics at AVR 75 or 77; however, because it is a two-field AVR, AVR 12 takes up more storage space and does not look quite as good as a similar low-resolution, single-field AVR. This is because the second field in AVR 12 is not to gain better resolution but to be compatible with the finishing resolutions. It saves time at the last stage because all your graphics will already be at AVR 75 or 77. If you were using the Meridien board, you would want to offline your timecoded material at the 20:1, two-field resolution to mix with the 3:1 or 2:1 for the graphics. You cannot mix uncompressed resolution with any other resolutions; however, in versions that have the new Batch Import feature, you can re-import your graphics at any resolution and link them up to the sequence. Now that Batch Import is becoming more common, having your graphics and timecoded material at different resolutions is not as useful for streamlining online.

Many film projects are using the multicam resolutions because they take up less space than similar "s" resolutions. If you are using the "m" resolutions, make sure to turn off or disable your 3-D effects. Do this by quitting the Avid and relaunching while holding down the "F" and "X" keys. If you are going to be using "m" resolutions consistently over a long period, then you should disable the 3-D in the Console. Go to the Console and type "disable3D" and return. After you relaunch the application, your 3-D effects will be turned off until you go back to the Console and type "enable3D" and relaunch the application. Your sound will always be CD quality or higher no matter what resolution you choose, so there is no need for redigitizing there.

TOTAL CONFORM

Many nonlinear systems claim to be able to conform Avid sequences, but none do it quite as well as Avid. The integrated use of OMFI information in the sequence allows Avid to open a project from an offline system inside a Symphony. All of the keyframes are present, along with other detailed, important information required for the most efficient re-creation of the creative work done by the offline editor. The more specific and painstaking the offline editor was, the more precise to that vision the final version must be. Total Conform will be covered in much more detail in Chapter 12.

NECESSARY EQUIPMENT

Your Avid suite is now an online suite if you plan to finish there. By purchasing the highest resolutions and the fastest drives, you have not completed this transformation! Online suites have some important components that must be present. There must be a high-quality third monitor to view reference video. This must be an engineering-quality monitor. You need one that can be adjusted through standard adjustment procedures like using the blue bars setting. The monitor must hold that calibration in a stable fashion over time. This generally means more money for the monitor, but if you ever get into a dispute about color or brightness, this monitor is your absolute. It is the end of an argument and the last word on "what it really looks like." Don't put your faith in cheap equipment—this finished image is your reputation!

Some people like to have a consumer-type monitor in the room as a low-end reference to "see what it will look like at home." Be careful of this since it is very hard in NTSC (Never Twice the Same Color) to ever get certain colors to look exactly the same on both monitors. When your client insists that the yellow on the low-end Gold Star monitor must match your high-end Sony monitor, you have a frustrating no-win situation. This is why many online tape suites have only one color monitor and everything else is black and white.

Very precise color-measuring devices allow you to adjust monitors precisely. After such an adjustment, a video monitor should be left on all the time to minimize the drifting that occurs with warming up and cooling down. If you need to adjust a color monitor, always wait until it has been on for several minutes, longer for older monitors.

It is also nearly impossible to get the computer monitors to match up to the video monitors. They are two very different types of monitors and, even though you can adjust the computer monitors, you cannot adjust hue through hardware adjustments. Color-matching extensions give you more control than the front panel buttons. Many professional graphics programs include these gamma-adjusting extensions, but you will never get the monitors to match completely. Optical from Color Vision is a product designed specifically for calibrating multiple monitors and will definitely help match computer monitors and video monitors. Do not let the client pick a final color on the computer monitor. They must have final approval on the well-adjusted, carefully lit, high-end video monitor. And yes, it will look different at home.

You will need external waveform and vectorscopes as an independent reference for all signals. Use it to monitor input, output, and dubs. If you can set up a patch bay or router, everything should go through these "scopes."

More and more people are deciding that having a color corrector in the suite gives them the extra protection and ability to deal with sources that have difficult color problems. It also gives you an extra tool to make sure that sources shot in very different locations really match up when they are next to each other. Since there is no hue adjustment for PAL, component NTSC, and no adjustment at all for serial digital input, the color corrector gives you the level of control you occasionally need. It can also give you some illegal levels if you are not careful! If that black looks really rich and the shadows have that deep dramatic look you really want, check that the black level never goes too low. The new Advanced Color Correction Engine in Symphony 2.1 will change the way the industry does color correction. We will discuss that later in detail.

You need studio-quality speakers, and they must be mounted far enough apart so that you can critically listen for stereo separation. If you can afford it, get a good compressor/limiter. This takes your audio levels and gently compresses the loudest parts of the sound so that they don't distort. This means you can have your overall sound louder and not worry about the occasional spike in sound level. You can use it during both the digitizing and the output. Most Top 40 radio stations do this to everything they broadcast, which allows them an even volume level: loud. If you are sending your audio to an audio expert to finish, don't compress it. Leave the sound alone except to adjust levels and make sure nothing is distorted.

CHOOSING THE INPUT

It may seem obvious that you always want to choose the highest quality input possible, but it is not always obvious exactly how to go about making the right choice. Let's start at the worst quality choice possible: composite. Composite video has the color information and the luminance information all crammed into one antique signal. This signal has changed little since the early days of television. The equipment that generates, carries, amplifies, and broadcasts the composite signal has improved significantly, but not the signal itself. The PAL composite signal is better than the NTSC composite signal because it has more scan lines and the color is more stable, but both can still be improved. Even if you must use a composite signal, you can avoid digitizing into the composite inputs of the ABVB (Meridien composite input is decoded much better).

The main problem with composite is that it must be decoded to separate the color information from the luminance. The decoding process is hit or miss unless you have dedicated hardware doing the separating of the signal: a decoder. Separating luminance information from chroma in the composite signal is as difficult as if I told you the number 12 and you had to tell me the exact two numbers I used to get that number. And the two numbers change all the time. You probably would get it wrong occasionally, maybe even most of the time. That is how a decoder must figure out where the luminance stops and the chrominance begins in a signal where they are combined. Good decoders actually change how they decode over time and have expensive, adaptive comb filters. Frankly, good decoders do a better job of decoding than simpler luma/chroma separation methods. If you are relying on composite sources for online quality, it is well worth your money to invest in a composite-to-component decoder. Buy only the types of converters that have a minimal delay since some models use a frame buffer and put you out of sync by one frame. If you only need to use a composite source occasionally, you might be better off sending the composite signal through a component tape deck, like a Betacam SP. You can use the component output of the Betacam SP to go into the video board for digitizing.

Some decoders even have an input port for S-Video and, if possible, you should use the decoder for that, too. Take the S-Video output of Hi-8 or S-VHS and decode that to component for digitizing. S-Video, sometimes called Y/C, is a better quality signal than composite because it separates the luminance (Y) and the chrominance (C) information from the beginning. But the chroma is just like composite chroma and has all the same problems with crawling and bleeding. You have more bandwidth for the detail in the luminance part of the signal, and the chroma does not have to be separated, but it is still not as good as real Betacam component. Middle-range and low-end Betacam SP decks have an S-Video in and out. As with the composite signal, you can run the S-Video signal through the Betacam SP deck and input component into the video board by using a Betacam SP as the decoder. Make sure that all tape decks are using the same blackburst signal for a reference to make this work. Again, the new Meridien video subsystem contains an excellent S-Video decoder, so don't be afraid to go directly into the input/output subsystem with an S-Video signal.

Then there is true Betacam component, more technically known as Y, R-Y, B-Y. The Y stands for the luminance and holds all the detail of the image. The R-Y is the red component of the image with all the luminance information eliminated, and B-Y is the same for blue. Together, they can create all the colors in the YRB color space (sometimes called YUV), what you can normally reproduce with Betacam SP images. The Meridien board accepts composite, S-Video, component (Y, R-Y, B-Y), and SDI.

Here is the chart for the various video boards:

	Composite	S-Video	Component	SDI
NuVista+	yes	yes	RGB only	no
ABVB (betacam top)	yes	no	YUV	no
ABVB (SDI top)	no	no	no	yes
Meridien	yes	yes	YUV	yes

Notice that the NuVista+ uses only RGB component video. RGB is the type of component signal the computer uses internally when it processes the signal. For the NuVista+ to accept Betacam SP component, it must be transcoded from Y, R-Y, B-Y to RGB. This requires a special box called a *transcoder* to change the signal from one type of component to another for both input and output. Getting a transcoder for the NuVista+ is really worth the money if you are finishing.

When you get into the digital formats like D1, D5, and Digital Betacam, the best way to go into Avid is serial digital (SDI). For this you need a special serial digital ABVB. Serial digital is also a good way to input DVC Pro since it also has a serial digital output. You may see other alternatives for DVC Pro in the near future that will take advantage of the faster than real-time transfer rates, but it will probably happen in news first. Serial digital input means that you don't have to convert from digital to analog and then back to digital again when working with digital tape sources. The digital information is copied over to the Avid just as if copying digital information between two hard drives. This produces superior quality, but if you have a mix of sources, like the occasional source tape from Betacam SP, then with the ABVB with digital input, you must convert from analog to serial digital with a stand-alone hardware converter. There is only one kind of video input with the serial digital ABVB, serial digital. When you output from a serial digital ABVB there is a monitor-quality composite output besides the serial digital output. This composite output should be used only for analog video monitors. To truly see how good the signal looks, you should be using a digital monitor anyway. You can probably use the composite output for a VHS preview, but not for finished output. Why waste all that quality? If you are mixing many different types of source for input, consider upgrading to the Meridien system and avoid all the confusion.

There is an issue with serial digital input to a system that uses compressed resolutions. Occasionally, you will need to work with very noisy signals either because of low light conditions or a very grainy film transfer. Keep in mind that the SDI retains all of this noise as well as all of the picture quality, which directly affects compression. It is much harder to compress a noisy signal, and you may find that a noisy signal through a straight SDI looks softer than component input when compressed. In this case you should consider taking the signal out of the analog component outputs of the deck and running it through an analog-to-digital converter for your SDI ABVB. If you have the Meridien board, use the component inputs to the video subsystem. Ironically, by reducing the amount of noise and the amount of complexity in the image, you may make your compressed image appear sharper.

With the release of versions 7.x/2.x, the Media Composer and the Avid Xpress with ABVB are compatible with EditCam media files, the field camera developed with Ikegami to record directly to disk. This new technology is exciting, and the idea of not having to digitize at all has obvious benefits in both time and money. You will also be able to add comments, do rough cuts, create bins, and perform many other traditional post-production tasks while still in the field. Connect the EditCam fieldpak to the editing system and import the clip information into your project. The media is already there and ready to edit.

SETTING LEVELS

In a perfect world, no tape should be digitized without being set up to broadcast specifications. This means looking at the tape carefully, both through the Avid internal waveform and vectorscopes, and external scopes. Both types of scopes serve important purposes and both are necessary. Don't set up an online suite without scopes! You use them when you digitize and when you are adjusting color effects, imported graphics, and dubbing. Getting a show rejected by broadcast engineers or a laser disc facility because the levels are outside standard limitations is costly and embarrassing.

Video levels with the ABVB and Meridien go from .74 IRE (not 7.5 IRE) to 108.4 IRE in NTSC and .249 volts to 1.063 volts in PAL. Note that this range is above and below the levels for legal broadcast, which is for a very good reason. The ends of this range are called *footroom* and *headroom.* Footroom is important for graphics using superblack (levels below normal black). Headroom helps to keep the signal from being clipped off if it is just a little outside of broadcast specifications. Every end use has its own set of standards. Public broadcasting in the U.S. has slightly more conservative standards than other broadcast outlets. Laser disk manufacturers have different specifications and may require different kinds and amounts of color bars and tone on the master tape. Find out what your requirements are and use the scopes throughout the project to conform to those specifications.

Many people memorize which scan line to use for each of the level adjustments using the Avid video tool, but this method will fail you the first time you are faced with different kinds of color bars. Instead, understand why you are looking at a specific part of the screen and what information the color bars are showing you there. In order to see the perfect color bar pattern, make sure the Digitize window is open, then open the waveform or vectorscope and hold down the Option (Mac) or Shift (NT) keys (see Figure 9.2).

There are several different kinds of color bars depending on where the tape originated. The most common color bars for NTSC purposes are SMPTE bars, which have several different sections or fields for different adjustments. The top section is used to adjust color on the vectorscope. The very bottom section is used to adjust brightness since it contains 100 IRE white. It is also used to adjust black level, or setup, because it contains three kinds of black, 3.5 IRE, 7.5 IRE, and

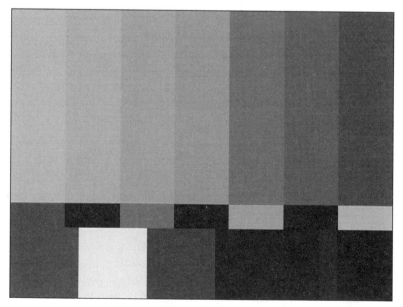

Figure 9.1 SMPTE Color Bars

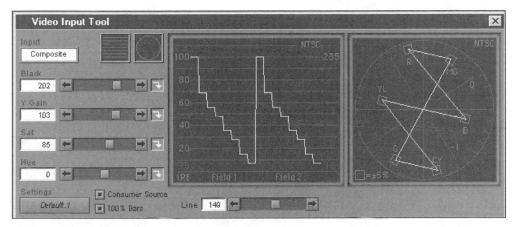

Figure 9.2 Holding Down the Option (Mac) or Shift (NT) Key with Digitize Window Open Shows 75 Percent Bars

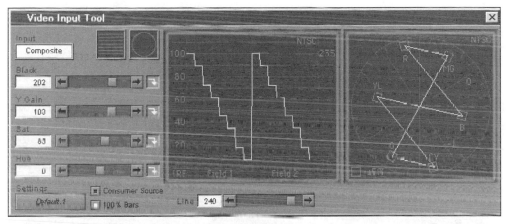

Figure 9.3 The Same Keys Show 100 Percent Bars When that Option Is Enabled

11.5 IRE. The middle value should be on the dotted 7.5 line. The other two values of black in NTSC are called *pluges* (Picture Line Up Generating Equipment) and are meant to be used to adjust your video monitor brightness. When you adjust brightness level on the monitor, you should see no difference between the pluge at 3.5 IRE (the one on the left of the group) and the setup at 7.5 IRE (the middle bar). PAL productions don't usually use color bars with a pluge. You can set the monitor to a cross pulse or underscan setting and compare the black of the now visible blanking signal to the black in the color bars. When you are adjusting the video input levels for gain (brightness) and setup, you must move the scan line being monitored to the bottom of the frame until it looks like this on the waveform (see Figure 9.4).

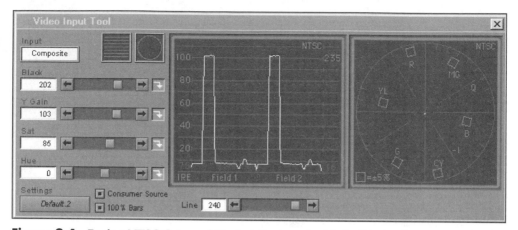

Figure 9.4   Perfect NTSC Gain and Black Adjustment with the Line at the Bottom of the Image

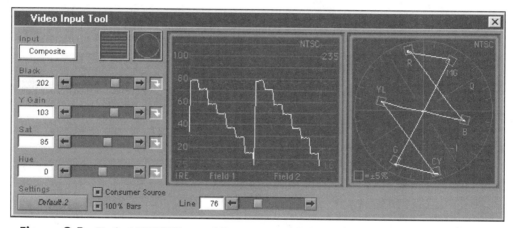

Figure 9.5   Perfect NTSC Hue and Saturation with the Line in the Middle of the Image

The second most common type of color bars are full field bars, which just have straight bars from top to bottom. These are simpler to use, but they tell you less. You can put the scan line just about anywhere in the image (except the very top and bottom) to get a proper reading for brightness and color. The important thing to determine about full field bars is whether the white bar is 100 percent or 75 percent white and whether the chroma saturation is 100 percent or 75 percent. Preferably, you should use the full field bars that have 100 percent white and 75 percent chroma saturation. These are commonly referred to, rather vaguely, as 75 percent bars and more specifically as 100/75 bars or 100/0/75/0 bars. If the white bar is at 100 percent, then set the white level to 100 IRE (NTSC) or 1 volt (PAL). If the white bar is 75 percent white, then in NTSC set the white level at the

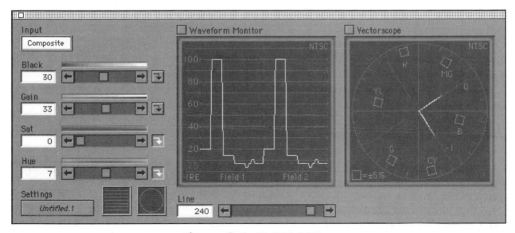

Figure 9.6 Full Field Bars

dotted line at 77 IRE. The math is strange in NTSC because of the 7.5 IRE black level, so 75 percent becomes 77 IRE. PAL has no setup, so 75 percent white should be at .75 volts.

There is yet a further complication and variation with the common 100/75 bars, which is the 100/100 bars. 100/100 color bars are not used often in NTSC because they cause composite video levels to rise to 131 IRE, and anything over 120 IRE would be clipped by the transmitter if it wasn't rejected immediately by the broadcaster; however, these 100 percent saturation color bars are quite popular in parts of the world that use PAL. This causes a problem with ABVB Media Composer Video Tool Settings, but it is handled with an optional checkbox on the Meridien. Most waveform and vectorscopes that can deal with 100 percent saturation actually allow you to change the way the scale is drawn to accommodate the higher level of chroma, which is what this new 100 percent checkbox does. I would strongly recommend against digitizing 131 IRE through the composite input of the ABVB.

In NTSC there is hue adjustment with a composite signal. You can move the color wheel around until the points of the vector signal line up in the little boxes of the vectorscope. With component input, you can only have more of B-Y (blue minus luminance) or R-Y (red minus luminance). More or less of these colors puts the vector points in the boxes. If the points don't fit in the boxes, you need to use a component color corrector during the digitizing or apply a color correction to the clip after it has been digitized. You can just set everything to the factory-set default levels of the video boards and let the Symphony online editor handle the problems! Actually, if you clip the signal by letting it get too bright or crush the signal by letting it get too dark, any color correction afterward will only mask the damage. Get the levels right the first time, and any further manipulation will look better.

A serial digital signal has no adjustments for video levels within the Avid Video Tool. You can adjust the levels on the tape deck itself or use a color correction after the image has been digitized. One popular digital color corrector has a Make Legal button or uses the Symphony SafeColor limiter. If only the rest of life was that easy!

WHERE TO ADJUST LEVELS

There are several schools of thought about the best place to set input levels: the deck TBC or the Avid. Frankly, according to the engineers who designed the ABVB, they are both good, with certain caveats. Let's look at both. You can decide which one is better for the way you work.

Output

Both methods require first setting the output levels of the ABVB or Meridien. The ABVB and Meridien have a separate tool for setting output levels, but the NuVista+ does not. If you do not set the output levels before you set the input levels, you may be masking a problem. In audio, if you turn down the volume of a signal that is already distorted, you get quiet distortion. In video, you get a darker image that has already lost all its detail. So make sure that the output levels are not affecting the signal at all. There is a very easy way to do this, but it requires using external scopes and a built-in test signal. If you (foolishly) do not own external scopes, make sure the levels in the Output Tool are in the green, preset position. Otherwise, use the toggle arrows at the bottom of the Output Tool ("More Choices" in early versions) and choose one of the color bar patterns under Test Patterns. These are uncompressed PICT files that are designed to show digital video levels (ITU-R.bt.601) and should register as the most beautiful and perfect pattern you have ever seen on your waveform and vectorscopes.

When you are outputting this perfect signal and looking at the external scopes, you may occasionally see a slight discrepancy. This may be because of other devices that are in the signal path between the video board output and the scope. You can eliminate this guesswork by putting the scopes immediately at the output of the video board. If this looks good, then you have a problem down the line, perhaps with a distribution amplifier (DA). A DA is meant to keep the signal from fading over long lengths of cable or to allow you to split the signal up into many outputs for routers or multiple decks. The DA must be adjusted as well!

If the signal is still a little off and you have eliminated the obvious, then you may have to adjust the output levels. If you do this, save them as settings, put them in the Site Settings, and call them "Default." That way, any new project will have access to these settings and can use them instead of the factory presets.

Input

The first method makes the most sense from the point of view of editors who have been setting all external devices to unity. In other words, before they send a signal to the DVE (digital video effects device), they adjust the input levels of the DVE so that all signals coming in look exactly the same coming out. That is called *unity.* What goes in is exactly what comes out. This method is also good for editors used to working with a complicated video pathway through a large online suite, routing system, and machine room.

Now feed the Avid a color bar signal, preferably from a color bar generator, but any tape that is known to have good color bars will do. You may want to make a tape of just color bars and 1 kHz audio tone and keep it around as your reference tape. The Avid outputs perfect color bars and tone from the Output Tool and the Audio Tool, and you can just roll tape for five or ten minutes and record it. Set the output of any playback deck so that it looks perfect on the external scopes and then feed the signal to the Avid. When you feed the bars to the input of the video board and set the levels so they are perfect on Avid's internal scopes, they should look perfect on the external scopes on the output of the Avid, too. Save these tape settings and call them Default. Save these settings somewhere safe and drag them to your Site Settings. You now have standard default settings for both input and output. This is what should be used with all tapes when digitizing.

Now you adjust the output levels of the video deck using the time base corrector (TBC) to adjust for the differences in levels between tapes. You will never touch the settings on the Avid. This way you know the Avid is not introducing any "noise" or affecting the signal level in any way.

There is one big drawback to this method: There is no way to save the settings on the deck while there is a place to save the settings on the Avid. Some TBCs have a memory setting, but it is up to you to recall them by hand when you load each tape. What if you want to use the same tape two weeks later and you want the shot to exactly match the same shot digitized at the beginning of the project? You had better hope that you set everything perfectly up to bars and then did not make any further adjustments because bars at the head of the tape are the only reference you have. Chances are, though, that you made minor adjustments after you set up the deck to bars because the picture didn't look just right. Unless you can reproduce that little level tweak by eye, you will have mismatching shots. It is much better to use the ability to save those levels and call them up automatically, and you can if you use a different method.

If you have not blown out or crushed a video signal while inputting, you can correct for any little adjustment in the Avid Video Tool without adding any extra noise to the signal. This is because the signal is digital, and as long as it stays within a particular range (not distorted), it can be moved around without any further degradation. In this scenario, you want the deck TBC to be always at the preset level. Throw all the video level switches on the deck into preset and then make the adjustments with the Avid Video Tool. Save the level settings based on tape name. Then, when you batch digitize, the settings for the tape are recalled automatically. If you have 100 tapes, this is a lifesaver. You set the levels once during the low-resolution offline stage. When you batch digitize to a higher resolution, and are in a little more of a hurry, the levels are right on without any further intervention by the digitizer. This method works as long as both the offline and the online are done on the same video board.

If you are adjusting the deck, look into a remote control panel for the deck's TBC controls so you can adjust quickly on the fly. You really can't count on the settings for the color bars being the absolute perfect setting for everything on the tape. So even if you are working from the Avid setting, you may still want to tweak the deck TBC occasionally. The problem with taking a deck's TBC out of preset is that it jumps to wherever the knob was set the last time you used it. Then you have to quickly tweak the levels to get them back close to the preset setting. This makes it hard to do just a little adjustment, for instance, while the deck is prerolling and you see something that is just a little off. A good method, although it is an extra step, is to put the deck in preset, set up the Avid Video Tool Input levels, and save them. Then, while watching the external scopes, take the deck out of preset and manually adjust the levels on the TBC so that they look exactly like they are in preset. You are using the power of the reference point — preset — to get everything right, then allowing yourself to make minor adjustments based on those original, perfect levels.

VERTICAL SHIFTING

Sometimes the entire image can shift down vertically and show you a black bar at the top of the frame and flashing little white lines. What you are seeing is VITC (vertical interval timecode). This is a second type of timecode that is recorded as part of the video. VITC should be on a video scan line just above the visible image. You will see this problem quickly if you have set your video monitor on underscan. The source and record monitors in the Avid show you this after you have digitized, but ideally, you want to catch this problem right away because it is so easy to fix.

This alternate type of timecode, VITC, is useful when you are moving the tape very slowly or very fast and the main form of timecode, LTC (longitudinal timecode), doesn't work as accurately because it is an audio signal. You know what happens to audio when you slow it way down; it becomes difficult to understand, and LTC is no different. Instead of guessing for the right timecode number at slow speeds, interpolating from the last well-read bit of timecode, a deck controller looks for the VITC. VITC also allows you to record extra pieces of information like tape name or film information that can be read by some timecode readers, like Media Reader, and even burned-in to the image while digitizing.

VITC is helpful until someone sets his or her record deck wrong and the LTC and VITC have different numbers. You see this problem if you are reading the correct number when the tape is playing and a ridiculous number when the tape is in pause. In that case, you want to go under the front panel of the video deck and change the switch to force it to read LTC Only. Imagine what this would do to your logs if you didn't catch it!

Back to the vertical-shifting problem: What is happening is that the video deck is being put into an Edit Mode in order for Avid to control it. When a deck goes into an Edit Mode, it changes where it looks for a reference signal for its internal timing. The video deck changes from using the blackburst it is usually being fed from the blackburst generator for reference. Instead, the deck looks to the incoming video signal. Many people are not feeding the deck a stable signal to the video inputs when they are digitizing. If there is no stable signal, it shifts the playback of the picture down several scan lines and reveals the otherwise hidden VITC. To prevent this situation, make sure the deck does not really go into the Edit Mode, always provide a stable video input to the deck (like a blackburst generator), or use a playback-only video deck.

The high-end series of record decks have a record inhibit switch that keeps the red light above the Record button on the deck always on. This is an easy and fast answer until you want to output a digital cut. If you don't turn the record inhibit switch back off, Avid gives you an error as it attempts to cue up and make the edit. It may even ask you if your deck is in the Record Mode! Another answer takes a bit more preparation: Make sure that every tape you put into the deck has its record inhibit tab in the inhibit position. Betacam SP tapes and higher formats have a little red switch or a screw on the bottom of the tape cassette. Three-quarter-inch tapes require a little red plastic tab to be filling a hole in the bottom of the cassette. If you don't have the plastic tab for the $^3/4$-inch tapes to fill the hole, then, as with most things in life, gaffer's tape does the trick. Open-reel formats, like 1-inch, have no protection on the tape itself, so you need to rely on the record inhibit switch on the deck. Confirming that the record inhibit tab is in the inhibit position should be done by every tape operator, every assistant, and every logger in the world as standard operating procedure. A quick elbow in the wrong place with an unprotected tape in a record deck and you have a big ugly hole in your master source tape. Try explaining that to the producer!

Once you have inhibited the video deck from being able to make an edit on the source tape, the image correctly repositions itself vertically. If this does not work, then you have a more serious blackburst problem and you need to evaluate that all pieces of equipment in the room are getting the same blackburst reference from the same blackburst generator. If you are working with a NuVista+ and a transcoder, this is probably the culprit. The blackburst reference input to the NuVista+ video board is the composite video in. If you are working almost exclusively with component or S-Video signals with a NuVista+, don't forget to also feed the composite video input with a blackburst signal.

This is an exhaustive look at setting levels, but they must be right. This is the biggest obstacle for many editors who have never been required before to care. Look at levels at every stage of the finishing process. What does that imported graphic really look like? I'll bet the print graphics designer didn't take into account broadcast chroma levels when he or she designed the logo. Train your eye to see color on your carefully adjusted monitor and scan for reds, cyans, and yellows that look too hot and double-check them through the scopes. You will learn what to look for, and you will feel much more comfortable with the whole process.

AUDIO LEVELS

Input and output audio levels must be analyzed before you do anything with audio. Generate a 1 kHz audio tone at 0 dB analog and input it to the Avid. There really is no industry-wide agreement on reference level for the digital level for audio reference tone, and it may vary from –20 dB to –12 dB. –20 dB is a new emerging standard audio used in video production. Digital video decks are calibrated straight from the factory at this level. Many engineers in the past have immediately adjusted everything that came from Avid to the –20 dB standard.

The digital reference level is adjustable in the Audio Tool under the right-hand pull-down menu (PH), Set Reference Level, and the scale on the Audio Tool will change. This is also now a Site Setting, so it will stay set to the –20, –18 or –14 dB. If you have an older system, and need to be compatible with a newer system I would strongly suggest that you recalibrate your Digidesign hardware to –20 by using the methods we will outline in the next section.

Playback the videotape with the 1 kHz tone at the proper level by adjusting the audio playback controls. If you do not have these on your deck, you need to double-check that anything recorded at the reference level of 0 dB analog also plays back at that level. Otherwise, your deck needs adjusting first. Then you must decide if you are going to use a mixer during the input stage. If you have many different audio sources like DAT, cassette, CD, or microphone, it might be better to run everything through the mixer. This also gives you a place to equalize or affect the audio one shot or one tape at a time. Now you must calibrate the levels through the mixer. Make sure that the 0 dB tone from tape is at 0 dB in the mixer when both the individual faders and the master faders are in the detent or notched position. Many mixers have trim pots for each input—little knobs in the back where each source can be adjusted individually as they come into the mixer. Be sure to pan odd-numbered channels to the left and even-numbered channels to the right.

Finally, look at the Audio Tool. Is it at 0 dB, too? If not, you have two choices depending on the hardware of your system. Whatever you do, do not touch the slider in the Audio Tool with the speaker icon. This adjusts the monitoring level, not the input level. This should always, always be at 0 unless you are specifically adjusting audio output that cannot be raised by any other means.

If you have a system that allows only two channels of input at a time, you either have the Audio Media II (NuBus) or Audio Media III (PCI) cards. These have an adjustment for input already in the Audio Tool that has a microphone icon. Bring the signal to the 0 dB analog level in the Audio Tool with this input slider.

If you can input four channels of audio at the same time, you have the SA4 (NuBus) or Amadeus (PCI) audio cards. Eight-channel input is standard on many systems with the 8×8 from Digidesign. This is a little trickier since it involves adjusting the levels by tweaking the Digidesign 4×4 or 8×8 that came with this system. These input and output levels may need some adjustment to be precise. The benefit to adjusting the Digidesign 8×8 is that once the hardware levels are set, no one should touch them again. The Digidesign hardware requires a small screwdriver, sometimes called a *tweaker* or *greenie* because it is used only for making these little tweaks and is usually green plastic. The larger Avid screwdrivers used for assembling the system are just a little too big and may strip the screw adjustment if you are not extremely careful. Carefully tweak the input levels until there are no yellow LEDs lit in the Audio Tool when inputting the reference tone. For a quick look at where the LEDs should be, call up the reference tone generator in the Audio Tool and observe where the perfect tone should be. Then, with the levels adjusted perfectly, digitize a minute or two of this tone into the system

Playback the digitized tone from the Avid and see if it plays exactly the same way it was digitized. It should. Now stop that playback and call up the reference tone again from the Audio Tool. Where does your audio go out of the Avid? If it goes through any distribution amplifiers, then these amplifiers should be checked out for unity. Then, does the signal go to the mixer again or directly to the deck? My vote is for it to go to the deck, but many people like to ride the levels one more time as the show goes to tape by putting it back through the mixer. This is fine, but it adds another level of complexity. It also means that there will never be two identical digital cuts of the same sequence unless the levels in the mixer are not touched. My approach requires getting it right in the Avid and then outputting as simply as possible, direct to the deck, to avoid those inevitable 2 A.M. mistakes.

If the output of the tone goes through the mixer, then put all the mixer faders to their notched position again and look at the level. Is it 0 dB? If not, adjust the trim pots for Avid inputs to the mixer. If you are not going through the mixer, then look directly at the deck. Set the deck record levels so that they record the tone at a perfect 0 dB (or the proper digital equivalent if you have a digital deck). Here, if you have the Digidesign 8:8:8 and are going directly to the deck, you can cheat a little if you want. If you have a separate tone generator, you can input directly to the deck and set the record level. Then see if the level coming out of the Pro Tools 8:8:8 matches this. If not, you may want to adjust the output levels of the Pro Tools. Alternately, if you trust the deck preset recording level, adjust the output levels of the Pro Tools to be at 0 dB when the input setting of the deck is in preset.

There is a twist that bears repeating if you are using the Audio Media II or III card. These two cards work with unbalanced audio and so must use cables that go from XLR to RCA connectors. They must also work with the input at –10 dB lower than the output. This is taken care of if you use the Avid cables that shipped with the system. If you must use other cables for some reason, then make sure they have a –10 dB attenuator or pad on the input to the audio card or you will never get the input and output levels to match.

Let's go back to our digitized tone already on the Avid and play back tracks 1 and 2 only. Look at the levels on the record deck and note where they are; they should be 0 dB. Go to the Audio Mix window and pan both tracks 1 and 2 to the right. How much does the level rise? About 6 dB. Now turn off the monitor for channel 2. There is a drop of 6 dB on the record deck levels. Keep this in mind as you mix your tracks. As you add audio tracks to the mix and then mix them down, the overall level gets louder; this is basic physics. If the stereo audio comes from a digital source like a CD and the mono sound comes from an analog source, you may have to drop the digital source level because it will have more headroom. The loud parts get louder than an analog source is capable of reproducing. If you have a sound compressor, you may want to compress the CD music so it is more even in level and does not get too soft or too loud in comparison to other analog sources. Some CD library music is already compressed for this reason.

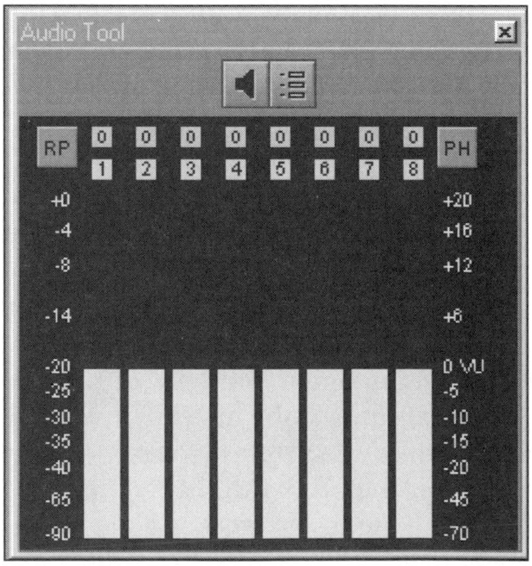

Figure 9.7 Audio Tone Calibrated to –20 dB

Changing the Calibration from –14 dB to –20 dB

Many facilities are already changing from –14 dB to –20dB so that they will be better calibrated with their digital video decks. You will increase the headroom but lower the signal-to-noise ratio from the published Avid specifications by plus or minus 6 dB. Most people will never hear the difference, and you will be rewarded by having fewer problems interfacing with these digital decks.

Before you begin changing the calibration of the hardware, you must make sure your software is also set to the right calibration. With the latest versions on NT (and soon Mac), you can go to the Audio Project Setting and change the calibration from –14 dB to either –18 dB or –20 dB (some countries prefer the –18 dB as their standard). The Audio Project Setting is a Site Setting and will automatically be used for every new project created on this system from now on. If you have an older system, you will not have the Audio Project Setting and must go directly to the Audio Tool to change the reference level and the calibration level. Then you must save the standard Audio Setting as a Site Setting yourself by copying the setting to the Site Setting window (under Special pull-down menu).

After you have made sure your software is set correctly, here is how you calibrate the hardware:

- Input a 1 kHz tone, +4 dBu @ 0 VU to audio channels of the Digidesign Audio 8:8:8.
- Open the Audio Tool and click the In/Out toggle buttons for all the channels to display "I" for input. The tone will appear as green level units.
- Click on the PH (Peak Hold) pop-up menu in the Audio Tool and choose Calibrate. This changes the scale to make it easier to be precise.
- Carefully insert a small screwdriver into the input trim pots on the outside of the Digidesign box. Adjust the level until the green level indicators of the Audio Tool show 0 VU.
- Do this for each channel until all eight are calibrated.

You must calibrate the input channels first, before calibrating the output channels. This is because you will actually patch the outputs back into the system to check their level and must rely on the input settings to be correct.

- Take output channels 1 and 2 and connect the audio cables to input channels 7 and 8.
- Click the same I/O toggle icons to display "I" for the channels 7 and 8 since we are using them for input right now. Toggle the I/O icons to display "O" for channels 1 and 2, which are now being calibrated for the output.
- Choose Calibrate from the PH pop-up menu.

- Under the PH menu, choose Play Calibration Tone.
- While watching the green level indicators in the Audio Tool for channels 7 and 8, adjust the trim pots on the Digidesign box output channels (1 and 2) and bring it to 0 VU.

Perform this calibration for all the other channels. When you are done, leave them alone!

TIMECODE ON THE MASTER TAPE

When you are ready for a digital cut, you need to consider how the timecode on the master tape has been set. The default timecode for sequences is 1:00:00:00, so you should either black the tapes so that you have enough space for bars and tone and start the show at this sequence time or change the start time on the sequence. If you start blacking your tapes at 58:00:00 or 58:30:00, then you have enough room for a minute of color bars and tone starting at 58:45:00 and then 15 seconds of black or countdown. (Actually, you have 13 seconds since there should never be anything in the last two seconds before the program starts.) Some countries use the 10:00:00:00 as the start time for their program on the master tape. To black tape in these countries, start at 9:58:00:00 or 9:58:30:00. If you are using the 10-hour start time, then you must change the Sequence Time (under General Settings) to start at 10 hours as well and save that in your Site Settings. If you do this correctly, then you can use the Sequence Time option during the digital cut. Sequence Time is preferable because there is a permanent relationship between the master tape and the sequence. This timecode relationship can be relied upon, even years later, if something has to change. You may also be able to edit into the master to make a small change without redigitizing the entire sequence.

For editors who are working with tapes blacked and timecoded with no regard for the Avid settings, you will have to change the sequence time of all the sequences to match the tape. Many editors leave their record decks in the preset timecode mode and whenever they do a crash record to start blacking the tape, the timecode starts at 00:00:00:00. By the time the bars, tone, and countdown are finished, the starting time of the sequence is around 00:01:30:00 or 00:02:00:00. You can change the starting time of all future sequences by also going to the General Settings and then saving the change as a Site Setting. General Settings are also the place where you change the default sequence time from non-drop-frame to drop-frame on NTSC projects. Highlight the number already in the setting and type a semicolon. The type of timecode on the master tape, drop or non-drop, must match the type of timecode of the sequence or you will get an error.

If you are putting multiple sequences on the same master tape, you may not want to take the time to change the start timecode on each sequence. In that case, there are the other two choices in recent versions Digital Cut window: Tape Time and Mark In. Tape Time is wherever the tape happens to be cued up at the moment; the tape backs up and prerolls to insert edit at that point. Mark In (in earlier versions, Custom Time) allows you to type timecodes in the deck controller at, say, regular intervals of two minutes. If you are using Custom Time in early versions, the type of timecode in the Custom Time window must match the type on the master tape. You must type the semicolon when you are inputting the Custom Time number if you have a drop-frame master. Once you make the change, it stays that way until you quit the application.

ASSEMBLE EDITS

Assemble editing has long been a feature request, and the addition of this function has saved many a session from stopping just to black a tape. Using an assemble edit also saves many hours of tape deck record head wear and tear.

Why would you use assemble editing? If insert editing was the only choice, you could not perform a Digital Cut until your one-hour show had a prepared black and encoded one-hour tape (timecode laid down over the entire tape while recording black). Maybe the plans change at the last minute and now there are two versions of the program, one with titles and one without. Do you have two blacked tapes? If not, don't black another tape before you begin—perform an assemble edit for the digital cut.

An assemble edit lays down new timecode during the edit. That timecode is determined by the settings on the record deck. There must be enough timecode on the tape for the preroll—a minimum of three to five seconds that you can set in the Deck Settings. But since the worst quality of any videotape is at the very beginning and the very end, give yourself a healthy 30 seconds if you can afford it. All professional videotapes actually contain well over two minutes more than the length on the outside box would imply. After blacking just that 30 seconds at the beginning of the tape, you can perform an assemble edit, and the digital cut lays down the new timecode during the playback of the program.

If you set the deck correctly to start the tape blacking at 58:00:00 or 58:30:00, then during the preroll for the assemble edit, the deck will read that code and regenerate it. The Deck Setting to start blacking the tape with correct timecode should be Preset, but then the switch must be changed to Regen before the assemble edit is performed. If you do not change the switch from Preset to Regen before the assemble edit digital cut, you will create timecode 00:00:00:00 at the edit point. Try to do another edit at that point and see what happens when you search backward from 0. The deck thinks it must rewind the 23 hours and 2 minutes to find the number for the inpoint, 58:00:00.

Where is the choice for assemble? If it is not in your Digital Cut window, then it must be turned on. Go to the Deck Preferences Setting and notice the checkbox for Enable Assemble Editing. Now you have a pull-down menu in the Digital Cut window with two choices — Insert and Assemble.

What happens if you do an assemble edit instead of an insert edit in the middle of the program? This is the most serious mistake you can make with a master tape, and it is the reason why the Assemble Edit function is not only just a little hidden, but it is also displayed in red (on the Mac). When the assemble edit is over, there are several seconds afterward where the control track of the tape is unstable as the recording comes to a stop. This can cause any image on that part of the tape to be unstable and unusable, and this cannot be fixed except by continuing the assemble edit to the end of the entire sequence. You cannot perform a clean insert edit over a point in the tape where the control track is unstable, like at the end of an assemble edit. You will always have a glitch. You should only use the Assemble Edit Mode if you are planning to lay the sequence down all the way to the end and there is nothing else on the rest of the tape.

OUTPUTTING TO NON-TIMECODED TAPE

If you are outputting to a non-timecoded master tape, you cannot control the record deck from the computer. This method is strictly for a quick crash record or going to VHS for a preview copy. Use the countdown built into the software to make the start of the presentation look a little cleaner and more professional. This countdown is purely for letting the tape come up to speed; it is not frame accurate and is slightly different each time you use it. If you want a frame accurate countdown, make one and edit it onto the front of the sequence. Spend a few minutes and change the PICT file for this countdown if you have a company logo. Any correctly sized PICT file for your video board and format will do, and it adds a nice touch.

COLOR BARS AND TONE ON THE MASTER

Where should you get the color bars and tone to put on the head of your master tape? One thing is for sure: bars and tone should be representative of the levels in the show that follows! Just sticking a pretty picture with lots of colors at the beginning of your show doesn't make the show look good. If you set a deck up by using these bars, will the show be legal, broadcastable, and listenable?

A few choices allow you to decide what works best for you. You can output both a perfect color bar signal and perfect 1 kHz tone directly from the computer as discussed before when setting all the levels in the edit suite. If you perform an insert edit using the edit controls on the front of your deck, you can lay both bars and tone to the head of the tape simultaneously. If you are confident that your video and audio that follow are just as perfect, this works out fine, with one

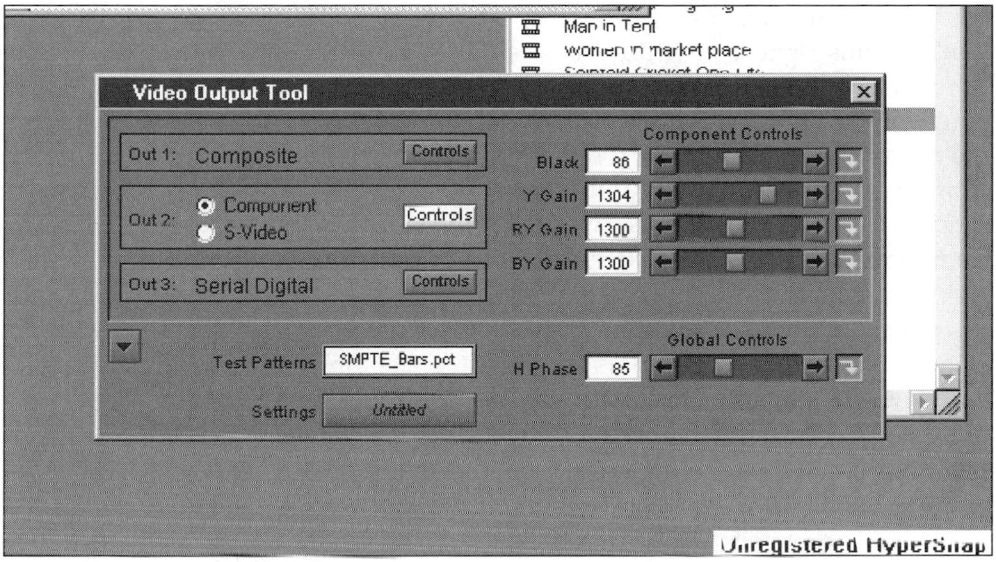

Figure 9.8   Output Tool Set to Show SMPTE Color Bars

exception. You want to make sure that there is no slight jump in black levels between the black on the master tape and the beginning of the fade-up at the start of the sequence. The little jump in black levels is visible to some people when they have the black levels turned up a little too high on their monitors at home. You could add two seconds of black filler to the start of your sequence and then change the start time of the sequence to 59:58:00. You can play back black filler from the Avid and use the front panel of the editing deck to insert edit onto the tape after the end of the color bars. Or you could splice all the bars, tone, and countdown to the beginning of the sequence and change the start time of the sequence to be 58:45:00.

IMPORTING COLOR BARS

The best way to get absolutely indisputable color bars and tone into the Avid is to import them. Avid systems ship with several kinds of color bars that are designed specifically for the Output Tool. If you call color bars up as the choice for a reference signal directly out of the Output Tool, you can adjust your Output Settings to be perfect. The files for SMPTE bars, full field 75-percent and 100-percent bars, are in the Test Patterns folder in the Supporting Files folder. These color bar PICT files are created at ITU-R.bt.601 values and are meant primarily for direct output only, not importing, but it can be done.

   If you import these files, you need to make sure that the import settings are set to ITU-R.bt.601 file color levels and aspect ratio, pixel aspect: ITU-R 601

(CCIR on some systems). Do not import them at RGB file color levels! If you don't do this right, you will have bars that are dangerously wrong! The white levels will be around 80 IRE and the blacks will be around 14 IRE. You will get your program rejected for broadcast if you import these test pattern bars at RGB file color levels. I have heard too many stories of the dubs looking wrong or the show being rejected, and the editor always claims, "These were the bars from the Avid!" Yes, but they were used wrong. Let me repeat: If you import the color bars from the Supporting Files folder, you must import with ITU-R 601 file color levels and use the ITU-R 601 aspect ratio.

There used to be another graphic that shipped with the software in the Goodies folder called Color Bars for Import. This graphic must be imported with RGB levels. Using RGB levels means that this graphic is for versions of the software and hardware that do not give you the choice of ITU-R.bt.601 values during import. There is nothing wrong with this graphic except that it is designed to be used in NuVista+ systems that cannot use any black level below 7.5 IRE. As discussed earlier, the split field SMPTE color bars have a pluge that descends to 3.5 IRE. This lowest pluge is not included in these bars. This doesn't affect the usefulness for setting up a signal; it means you have one less tool for adjusting the brightness of a monitor. If you have a NuVista+ system or your software version cannot import at ITU-R.bt.601, this is the only graphic you should import for correct video levels.

If you are confused or suspicious about what color bar levels you have, check the bars at the head of the sequence, either through the internal Avid waveform monitor or through an external waveform and vectorscope. If the color bars generated by the Output Tool look perfect, but the color bars at the head of your sequence do not, then the bars in the sequence must be replaced.

Audio tone must also represent the level of the audio to follow. Recent versions give you the ability to create a perfect tone media file in the Audio Tool. This can then be edited onto the front of the sequence with the imported color bars. All of this preparation can be done as a last step before performing the Digital Cut. Personally, I hate editing with a sequence that has all of this reference information at the head, so I put it off until the very last stage. When I use the Home button, I want the sequence to go to the first frame of the show, not color bars.

Another school of thought says that the color bars must be digitized through the video board if they are going to match the rest of the video that has also been digitized through the video board. It is much more important that the levels all be the same throughout the sequence and run through the same scopes. This argument also does not take into consideration that a substantial portion of your sequence may have been imported and not digitized at all! But just for convenience, you may find that carefully digitizing color bars, tone, your own special slate, and countdown and black is just easier for you to edit onto the head of your sequence. If the levels are not representative of the video that follows, this method is not any more accurate than the other methods.

THE FINAL CUT

When it is time to actually perform the Digital Cut and everything else has been prepared, there is one more setting to check. Note when you call up the Digital Cut window which tracks are highlighted. You can patch the audio or video layers from the Avid to the tracks on the tape, making it easier to customize the master tape without making multiple sequences. Maybe you are going to add titles in a last digital pass in a linear suite because you have a specific look that can only be created by a Chyron character generator. You want to lay off a copy of the show without the video track that contains the titles. If an assistant will perform this function, he or she must be quite clear on the implications of this setting.

What if you are viewing the Digital Cut one last time as it goes to tape and discover a minor mistake? You can perform an insert edit in the middle of the sequence without laying down the entire sequence again. Make sure you have the type of edit set to insert in the Digital Cut window first. Then mark inpoints and outpoints in the sequence at existing cut points before and after the change. Then uncheck the Entire Sequence checkbox in the Digital Cut window and highlight only the tracks that need to be changed.

If the change lengthens or shortens the sequence, be aware that there will be some more work to do at the end. If the sequence is now shorter after the change than it was the first time you recorded it to this tape, then you must insert black and silence onto the tape after the sequence is over. There is almost nothing more embarrassing than fading to black and then coming back for a little reprise of the last few seconds! You can add black to the end of the sequence, but unless you have the latest versions, it is not as easy as you might imagine. With the latest versions, there is a checkbox called "Add black at tail." You can determine the amount of black with a pull-down menu. This will not change the length of your sequence but will keep the deck recording for a few extra seconds after the sequence has stopped playing. Make sure this option is turned off if you are just replacing a shot or two and are not re-recording the master all the way to the end.

On older versions, you cannot just add filler to the end of a sequence in the Avid. There are a few workarounds to choose from if this is desired, but I prefer the simplest method. Use the edit controls on the front of the master deck and do a short insert edit while playing back filler from the Avid. This will not work if you don't have direct access to the deck or your deck does not have front panel edit controls, so you may be forced to use one of the following methods.

You can import a black graphic, or digitize black and edit it onto the end for the length needed. You can also splice any piece of audio on the end and then turn down the volume for that clip. Other methods to make the system think there should be filler at the end can backfire when you make later changes and the sequence drops the fake filler. This happens if you splice a piece of video on the end and then lift it out. This is fast, but only a temporary fix. Any black or sound added to the end of a sequence ends up in the EDL if you don't remove it, which is why I prefer the method that does not affect the actual sequence.

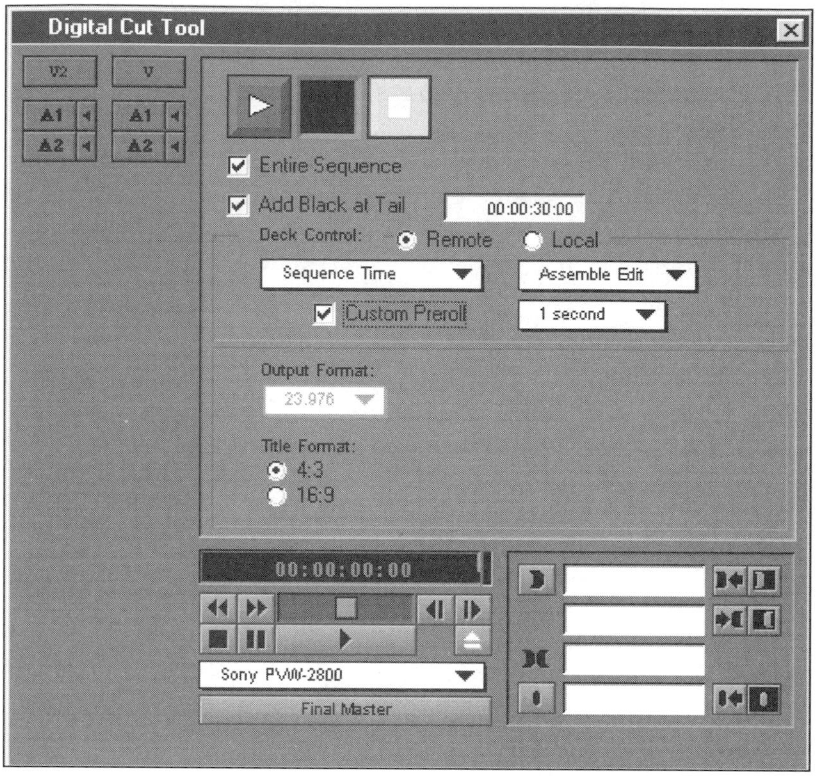

Figure 9.9 Digital Cut Tool Set to Assemble Edit

If the sequence is longer than the first pass after the changes, then make sure the tape has been blacked long enough to accommodate the extra time. If not, change from insert mode to assemble mode in the Digital Cut window, and on the record deck, put the timecode Preset/Regen switch in the Regen position.

CONCLUSION

This is an overview, although detailed, of all the various considerations to prepare before beginning to finish a project on the Avid. Once these issues are dealt with in a consistent and thoughtful way, then many of the annoying quality issues will be absorbed into standard procedures. The next two chapters will go into specific strategies and features designed to speed workflow and to help you cope with flexible and fast-changing situations. Once the basics have been prepared, much of this process can be streamlined to focus on the creative rather than the mechanical.

10

Offline to Online: Finishing Strategies

After you pick the right resolution for your project, the digitizing can begin. There are many ways to approach redigitizing depending on the length and complexity of the project. The more involved this last stage is, the more you will benefit from creating a strategy and sticking to it throughout the project. Making sure everyone in the process knows what the final result should be means that all carefully laid plans will still be effective in the final crunch time. This chapter will cover both simple and complex procedures that will allow you to get the maximum flexibility and power from your nonlinear finishing system.

The important factors in finishing are quality control, the speed of graphics replacement, and amount of rendering. The video and audio levels must be monitored at every step to make sure they conform to broadcast specifications. Most people render too much, so Avid fixed this problem with ExpertRender™. If you are trying to make an air date, then saving even a few minutes becomes crucial. If you are trying to complete a long and complicated project, then you may need these extra choices and procedures to get exactly what you need.

USING OFFLINE LEVELS IN ONLINE

It is important that you set the levels for each tape at the offline stage if you are planning to redigitize on the same system. You can save these Level Settings attached to the tape name, and they will be recalled automatically at the batch redigitizing stage of the online. These settings will only be applicable to the system you are working on now. If you save tape settings and use them on another system, that video board may be calibrated slightly differently. If you are moving to another system for high-resolution digitizing or to a Symphony system, you should delete all the Tape Settings from the Project window. If you can get the levels correct at the offline stage, you could save huge amounts of time later when the deadline is tight and time is potentially more expensive.

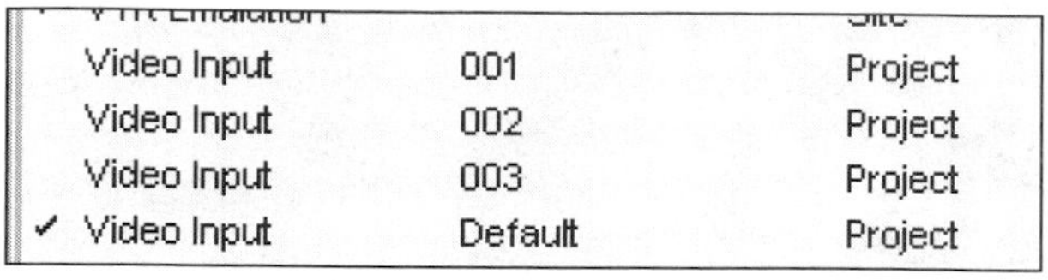

VTR Emulation		Site
Video Input	001	Project
Video Input	002	Project
Video Input	003	Project
✓ Video Input	Default	Project

Figure 10.1 Tape Settings Can Be Found in the Project Window

If there are multiple, radically different levels on the same tape, you may end up creating several settings for that tape. Give that tape multiple tape names — a new name for each setting. For instance, you may have tape 005, which is primarily exteriors. At the end of the tape there is a change to an interior that requires a different set of video levels to compensate. You might pop the tape out of the deck, reinsert it, and rename the tape 005IN or something similar. Change the Level Setting to compensate for the interior and label the outside of the tape to reflect that the system now thinks there are two tapes.

After the levels are set correctly for the video, make sure that the audio levels are set correctly, too, and watch them closely. Always keep in mind that you can use this audio as the finished mix. If you are going to end up on digital videotape, you can digitize at 48 kHz and go digitally to tape (AES/EBU) for the final output. If you choose 48 kHz now, you won't be able to move the project halfway through the job to an older system without the 48 kHz option. There are DAT decks that accept 44.1 kHz, so they are a good alternative if you have an older system and want to work with your digitized audio as a final audio mix.

The older, lower end Avid Xpress and Media Composers may input and output S/PDIF (Sony Phillips Digital Interface Format). S/PDIF may not be available on some professional-level DAT decks, so check the compatibility of your digital connections before starting the project. Many DAT decks use only the AES/EBU standard connection.

If the audio goes into the system right and you can mix the sound as well as needed in the Avid suite, then you will save a significant amount of money. A MIDI controller to ride audio levels in real time and the AudioSuite sound plug-ins are excellent tools for the finishing touches or special effects. On the other hand, professional audio designers can be worth their weight in gold on the right job. You may be able to hand those audio media files directly to the audio designers as AIFF or SD2 files with OMFI sequence information for Pro Tools, AudioVision, or other digital audio workstations (DAW). This will save you time and money at the beginning of the mix session.

When graphics and animations are created, you import them and cut them into the show as needed. The best scenario, especially for older systems, is to import graphics and animations at the finishing quality of AVR 77 or 2:1 (you can't mix 1:1 with any other resolution). If not, you will need to reimport them later when you get to the online stage. Reimporting is easy now with Batch Import on Meridien systems, even if the graphics have alpha channels, but this is rather tedious on older systems. If you use many graphics when you finish (especially with alpha channels), I would highly recommend upgrading since the time savings on just the first few jobs will make a big difference. If this is not an option, I will outline a method later in this chapter that I devised to batch import graphics and animations without alpha channels.

Edit the sequence together until you get the final authority to sign off. Many people call this *picture lock* and assume that nothing is going to change after everyone sees this final version. (Don't ever believe this, but it is a nice idea.) In the process of getting final sign-off, you need to get this low-resolution version as close to the final high-resolution version as possible. That entails, in many cases, creating all the effects at this offline stage. If straight cuts are more than enough for your production, consider yourself lucky and in the minority. You may have a specific look that has been mandated by the designers, and it must be accurate down to the kerning and leading of every letter of every title. This, too, must be created in the offline so that there are no chances for them to say, "I didn't know it was going to look like that!" and start redesigning while the clock ticks. Everyone's approval in the offline stage is important. Again, the Total Conform from offline to uncompressed online is one of the main values of using an Avid system like Symphony. This is a critical, competitive advantage at this stage.

SIMPLE REDIGITIZING

When it is finally time to take the final cut and redigitize at the finishing resolution, you have two choices of action. You can choose either to redigitize the entire length of the original master clips used in the sequence or to just digitize what is necessary to play the sequence. The first choice, digitize all of the original master clips, is simple but takes up much more disk space than being a bit more selective. To find the master clips used in the sequence:

- Put the sequence in its own bin.
- Under the pull-down menu in this new bin, choose Bin Display.
- Choose Show Reference Clips.

The master clips used in the sequence will appear in the bin. These clips are always connected to the sequence, but they are usually hidden to the user unless you ask to see them.

- Select all the master clips and batch digitize.

Perhaps you need to be more selective about what you actually digitize because you need to save time or disk space. Consider one of the many ways Avid allows you to be efficient during the batch.

Batch Digitizing a Sequence

Here is the simplest way to batch digitize a sequence:

- Take all of the low-resolution media offline. Do this by either deleting or moving media out of the MediaFiles folder.
- Change the resolution in the Compression Tool.
- Highlight the sequence.
- Choose Batch Digitize from the Bin menu.

First, the system asks if you want to batch redigitize everything or only the media "which is unavailable." Because you have taken all of the media offline, all of the media is unavailable. The system can keep track of what has been redigitized, so when you stop and start, it knows what has already been digitized (that media will be "available"). The system skips the available clips.

The system asks you for the required tapes and takes every shot it needs from each tape. The beauty is that the system only digitizes as much of the shot as it needs to play the sequence with a little extra thrown in as *handles*. The handles give you enough media to make minor changes, like the length of a dissolve, or allow you to slip the shot a few frames to avoid a dropout. Do not underestimate the importance of handles, but do not forget that you are using up disk

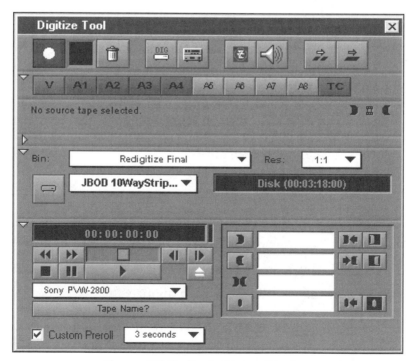

Figure 10.2 Batch Digitize Window

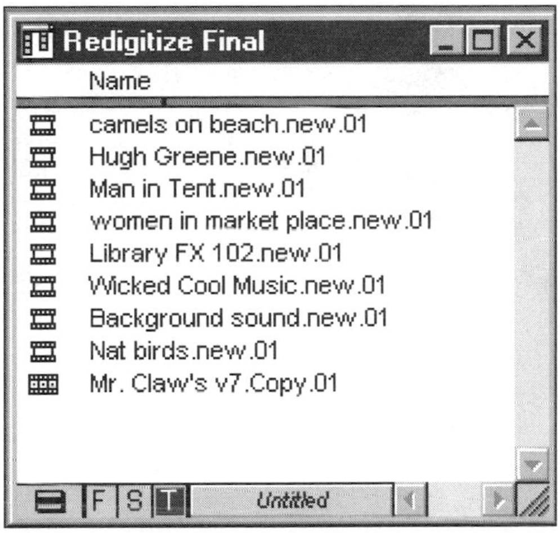

Figure 10.3 New Clips in a Redigitize Bin

space very quickly at the high resolution. At the end of the batch digitizing process, you will have a bin full of new master clips that have the original clip name plus ".new" and a number at the end. This is your new high-resolution media.

Batch digitizing the sequence this way is simple, but it doesn't give you very much control. If you have a very complex or long sequence, you are at the mercy of the system in terms of stopping or pausing or even trying to anticipate what the next needed tape will be. You might want to leave all of the low-resolution media online when you start redigitizing because you will continue to edit low-resolution versions after making the high-resolution version. If you do this, you cannot take advantage of the "digitize only media which is unavailable" feature. All shots will be "available" and the system does not distinguish between resolutions. If you stop digitizing and want to pick it up later, it is not so easy. You have to skip over every shot and every tape already digitized, or every tape not present, one shot or one tape at a time. If you have skipped tapes on purpose, maybe because they were not ready, you have to manually skip over them with "skip this tape." You will have to skip shots and tapes each time you stop and start again. With several hundred tapes, this can get old fast.

DECOMPOSE

Decompose is the feature you can use with Media Composer and Symphony to specifically address the need for redigitizing sequences. It is designed to give control back to you by breaking the sequence into new master clips before digitizing. The basic logic of the system tries to protect you from digitizing over

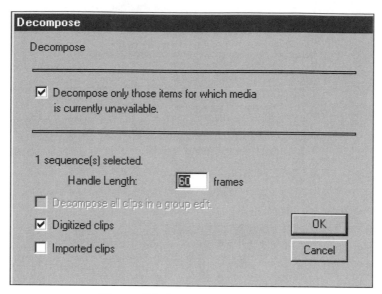

Figure 10.4 Decompose Dialog

material that is already there. This is why, in the first scenario, you should take media offline before you batch digitize. When you decompose, the low-resolution media is taken offline automatically as the first step before redigitizing.

Duplicate the sequence and put it into its own bin. Highlight the sequence and choose Decompose under the Clip menu. When you decompose, you get a warning that is a little intimidating about how this function will take all your media offline. This message means that, for this sequence only, it will break the links to the media already on the drives. In one step, it creates the new master clip with ".new" on the end. Since these new master clips have never been digitized before, they link to nothing on the media drive, so all media is considered "currently unavailable." There is no need to take any media offline before batch digitizing.

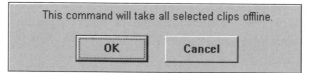

Figure 10.5 Decompose Unlinking Warning

Now you can arrange the clips in any order you like. You can easily skip over tapes that are not ready yet and sort or sift the new master clips in the bin. When you finish batch digitizing, you have all the high-resolution media you need to play the sequence plus handles for minor changes. We will come back to the Decompose option later in this chapter because it is so important as the first stage of some sophisticated variations on this simple redigitizing process.

All titles must be re-created, an automatic process that leaves them intact, but changes the resolution. If you have graphics or animation and you cannot import them at the finishing resolution the first time they are used, then you must reimport them now at the higher resolution. All imported graphics with or without alpha channels can be batch imported (or with older versions must be recut into place by eye). If you are using older versions, graphics and animations that have no alpha channels can be automatically relinked by a workaround method I will describe later. All effects must be rerendered, then everything is digital cut to videotape.

MORE COMPLICATED DIGITIZING SCENARIOS

Drive Space and Offline Media

Your strategy for redigitizing is going to first require a decision about drive space. Do you have enough space on your media drives to keep all of your low-resolution media? This is a critical decision since you may need to recut sections and make multiple versions after the final sequence is finished. Maybe your producer will want to make changes after viewing the final high-resolution master and you will have to go back to the source material to find new shots. Deleting your low-resolution media is a big step toward the point of no (quick) return. This should be done only after serious consideration. It also significantly impacts the way that you manage your media during the redigitizing. Let's look at several scenarios depending on whether you keep or discard media while streamlining and minimizing the redigitizing process.

Splitting the Sequence

In the first scenario, you have enough space for all the media to be on the drives at the same time. The decision is also to keep all of the audio exactly as it is right now. There has been lots of sound mixing or the mix will be done by someone else. Either way, there is really no reason to redigitize audio unless it is distorted or is being replaced with material that is much better quality, say the DAT original instead of a U-Matic dub.

If the audio needs to be redigitized, then the mix has to be tweaked because the levels during the redigitizing may be just slightly off what they were originally. You can save time and money if you can do anything to keep from redigitizing and remixing your sound. Think seriously about changing your offline methods so that the audio does not need to be redigitized during the finishing stage.

This strategy entails stripping the video away from the audio when starting the redigitizing process so that you are working only with video clips. There are several easy ways to do this, but let's use the easiest first.

- Duplicate the finished sequence and put it into its own Redigitizing bin.
- Load the sequence into the Record side of the Source/Record monitor.
- Highlight just the audio tracks, and hit Delete and OK.

You are left with only the video tracks.

- Highlight the same sequence in the bin.
- Choose Decompose from the Clip menu.
- Uncheck "Decompose only those items for which media is unavailable."

What this really means is, should this function create new master clips for everything in the sequence or just the media that is missing or offline? Since everything is available or on the media drives and linked to the sequence, then leaving this box checked results in nothing happening. This can be used effectively in other strategies, but not now. This is not the desired outcome, so uncheck this box and, whether the media is available or not, the Decompose function creates a new master clip for every shot in the sequence and adds ".new" to the end to make sure there is no confusion with the original (longer) master clip.

Using MediaMover to Take Video Offline

There is another way to take video offline that does not require splitting the original sequence into two separate sequences of audio and video. Chapter 4 discussed a third-party program called MediaMover. This program can search all your media disks and find all the media for all the projects used in this sequence. MediaMover versions 2.0 and later have the capability to move just the audio or the video out of the MediaFiles folder and into a separate folder with the project name. This effectively makes the media offline or unavailable. If you do this to the video-only clips, when you use the Decompose function, you can leave the "decompose only clips for which the media is unavailable" checkbox checked.

Decompose makes ".new" master clips for the video only. If you have used material from many different projects, then this method is not very effective since you will have to move the media for all the projects involved. Either way, splitting the video and audio tracks or moving the video out of the MediaFiles folder, you end up creating new, offline, video-only master clips.

Handles and New Master Clips

Decompose may not actually create a new master clip for each shot in the sequence. One exception is imported graphics and animations. These are not decomposed automatically because they cannot be batch digitized. On the recent versions, there is a checkbox to determine if you really want to decompose the imported graphics or animation. This is no longer a problem because Batch Import simplifies the entire process to one step.

Another reason you may not get a new master clip for every shot is the second part of the Decompose dialog: the handles. The handle length defaults to 60 frames and is easy to change. If you shorten it and you have hundreds of cuts in your sequence, you save yourself considerable disk space. One hundred clips times 120 frames per clip for handles add up to approximately an extra 6.5 minutes of footage in NTSC. If your one-hour show has 600 to 800 edits, this is a serious amount of extra storage space at high resolution. If you make the handles longer, then two shots in the sequence that were right next to each other on the source tape may actually overlap with timecodes. The Avid determines if there is an overlap and combines the two (or more) shots into one master clip and so reduces duplicated frames. Combining multiple shots intelligently into a single master clip reduces the amount of master clips and speeds up the redigitizing process by eliminating the amount of time needed to stop, search, and preroll for each shot individually. Extrapolated over hundreds of shots, this can really add up to a time savings.

Automatically Offline

When you decompose, the sequence breaks its links with the low-resolution material and links to the new offline master clips. The ".new" master clips are offline and they are video only. You could sort and sift, stop and restart the batch digitizing to a higher resolution and have complete control over the process, but there is one drawback: What if you are asked to continue to edit with these new uncompressed video-only clips? Suppose you are asked to make a new version of the sequence? Where are you going to get the sync sound? From the original audio? And then sync every shot by matching timecode? All of this seems like a huge amount of work. Working with video only can backfire and force you to lose the savings created by leaving the audio alone. If you are totally sure that you will not be asked to continue editing at all with this material, then go right ahead and start batch digitizing. But if you know your client too well, there may be an extra step.

You can modify the video-only master clips, add audio tracks to them, and continue to edit using sync sound. This seems totally redundant after you just stripped the audio out, but this new audio will be used for a different step in the process. You will not use this audio for the final mix! The levels do not have to be

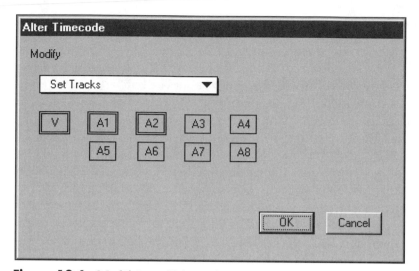

Figure 10.6 Modifying a Video-only Master Clip to Add Audio Tracks

re-created perfectly; the new audio is for reference only and you will not accidentally cause the sequence to link to this new audio. It is only used in case you are asked to go back to the high-resolution shot and play the whole thing through. If you match frame on the video track in the final sequence, the system calls up a shot with sync audio. If you match frame on an audio track in the sequence, you call up the original master clip with the original audio. You really have the best of both worlds with one minor drawback: Redigitizing the audio takes up more drive space. But audio really takes up very little space compared to video, and if your sequence is really not that long, the difference in storage is minimal.

Consider the extra amount of audio along with the extra amount for handles when redigitizing. One way to figure out how much space you need is to select all the decomposed master clips in the Decompose bin and hit Command-I or Control-I (Get Info). This calls up the Console and gives you the total length of all the clips in the bin. In earlier versions of the software, you need to open the Console first before using the Get Info.

The Tape Settings created during the offline are automatically recalled during this batch redigitizing process. You may not want the Tape Levels Setting to be recalled! Perhaps someone set the tape up incorrectly or you are using the original Betacam SP reel instead of the S-VHS dub and the two tapes are quite different looking because of a bad dub. You may be batch digitizing on a Meridien system, although the offline was done on an ABVB system. You have to create a new setting with the same tape name to replace the existing one or go to the Project window and delete the Tape Setting completely.

Some people make the deck preroll time a little longer during this stage so they can quickly correct any little level problem they see before the shot starts to

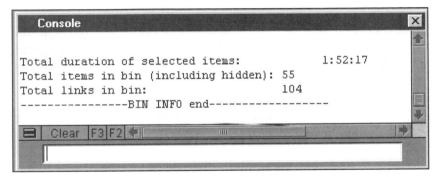

Figure 10.7 Get Bin Info in the Console Before Batch Digitizing

digitize. This is why you take the deck adjustment out of preset. Many times the shot requires a tiny raising or lowering of the video gain to even things out.

The tapes are requested by the system one by one and each tape is used only once. The system starts by digitizing the earliest shot on the tape. Then the tape moves forward to the end of the reel, digitizing all the shots that are needed. You can speed up this process by either rewinding all the tapes beforehand or actually cueing up to the first shot that is needed on that tape. This is easy to do if you have an extra deck or two and a printout of the decomposed bin.

Putting the Sequence Back Together

Once the high-resolution video and reference audio are digitized, then use that sequence as if it were a source clip. Overwrite the high-resolution video-only sequence as the new video track for the original sequence. The new video should match perfectly and have the same starting timecodes. If you hit the Home button and mark in on both sequences, the new high-resolution sequence will be in perfect sync with the original sequence. Even the sync breaks work because they are looking for tape name and timecode. So if you continue to tweak the sequence and you knock yourself out of sync, you are still given the white numbers in the Timeline. This may be the last step for a simple sequence.

The Low Disk Space Variation

When you have very limited disk space, then all low-resolution media must be deleted and the audio must be reduced to the bare minimum to free up space for the high-resolution material.

Here, you can use the Media Tool to delete large amounts of low-resolution video all at once. Choose to show all the media files and precomputes for this project. Select all the media in the Media Tool and hit Delete. A dialog box appears and asks you if you want to delete audio or video. Uncheck the audio choices and delete all the video only. Scan through the sequence to make sure all of the video is gone. If there are any shots left, then they were part of another project. You must decide whether they can be deleted or not. To be completely sure this method will work, you should either delete these leftover shots or figure out how to take them offline. You can do this by:

- Dismounting the media drive they are on.
- Moving the media from those projects out of the MediaFiles folder using MediaMover.
- Locating the media file using Reveal File and moving it out of the OMFI MediaFiles folder.

This is a variation of the first method and is similar except that you have deleted the project's video media instead of just moving it out of the way.

If you are still tight for space, you can consolidate audio used in the sequence. This takes the audio media and copies only the parts you need.

- Copy the sequence into a new bin.
- Select the sequence and choose Consolidate from the Clip menu.
- Uncheck "Skip if media is already on the drive" because you are not just moving the media, you are shortening the original master clips to only the parts you need.
- Check "Delete after consolidating" if you have set the Consolidate drive list to include enough drive space to finish the entire sequence.

On older versions if you are tight for space on any one drive:

- Split the sequence into pieces and consolidate each piece to a separate drive.
- Go back after the consolidation and delete the original audio media by hand; this method is slower but you have more control.

Now you have a sequence that is linked to the consolidated audio and you can decompose the video only. Because you deleted the video as the first step, it is already offline. Decompose "Only those items for which media is unavailable" and get video-only master clips to redigitize.

Mixing AVR 12 and AVR 77

This scenario involves mixing AVR 12 and AVR 77 material, and you want to keep all the AVR 12 material before redigitizing. Basil Pappas solved the problem with the best method I have seen. I keep this method for ABVB only, although it will work just as well with Meridien. With the Meridien releases, the Batch Import feature has made this method less valuable, but still handy as an advanced technique.

If you don't need the AVR 12 material, you can sift for it in the Media Tool and delete. Now you need another way to separate the AVR 12 from the AVR 77 because you want to keep all of your media available.

- Copy the sequence and put it into a new sequence bin.
- Strip the audio tracks from the duplicate sequence and make it a video-only sequence as in the previous technique.
- Open the Media Tool and show master clips from all projects.
- Make sure nothing is highlighted or selected in the Media Tool.
- Highlight the new video-only sequence and, in the new sequence bin, use the Bin menu choice, Find Media Relatives. This highlights all the clips in the Media Tool that are used in the sequence.
- Drag all the highlighted clips from the Media Tool into the new bin.
- Sift the clips in the new bin so that only the AVR 12 clips are showing.
- Select them all (Ctl/Command-A).

Now you are going to unlink these clips from their AVR 12 media and purposely make them go offline.

- Hold down the Control and Shift keys at the same time and under the Clip menu, choose Unlink.
- All the AVR 12 material unlinks and becomes unavailable.
- Move the video-only sequence into yet another new bin.
- Select the sequence and decompose. Check "Decompose only the clips for which media is unavailable."

The sequence will decompose only the video clips that were linked to the AVR 12 media. Decompose creates new master clips for all the original AVR 12 clips and leaves everything else in the sequence alone.

- Go back to the bin with all the unlinked AVR 12 clips.
- Select all, and relink them.
- Check "Relink master clips" in the Relink dialog. All the AVR 12 clips should go back to normal.
- Digitize the decomposed, video-only master clips that you just created.

If you like, you can also take the extra step and modify the video-only master clips and add audio tracks.

Decompose without Timecode

Many people digitize non-timecoded material at the finishing resolution during the offline stage. They know they can't redigitize it later in the traditional offline to online process, so they want to digitize it only once. This method works well if you have only a few shots without timecode and don't want to use the previous method to isolate them. What happens to the shots with no timecode when you decompose? If they were decomposed along with the rest of the sequence, then now they are offline in the sequence. How do you get them back?

When the decompose and batch digitizing process is finished, there will be gaps in the sequence. This is where the non-timecoded material should be. You can check to see if you have any gaps in your finished sequence by putting a copy of the redigitized sequence into another bin and decomposing the sequence again. This time leave the box checked for "Decompose only the clips for which media is currently unavailable." All the media should be available now except those shots not redigitized. If you missed a shot for some reason, it is a new decomposed master clip. This is used purely as a simple way to pinpoint what is missing; you don't really want to redigitize these clips. In later versions, a better method is to use Show Offline Media in the Timeline. This will give you a fast visual reference since all the offline shots in the sequence will be highlighted in bright red, even if they are nested several layers deep.

When non-timecoded material was originally digitized, it was given a time-of-day timecode that cannot be re-created. This timecode is made up by the system and assigned during the digitizing and is not really on the tapes. This fake timecode is enough so that the system thinks this material is different from a graphic, video mixdown, or another kind of imported media that has no timecode at all. You have a choice of whether to decompose imported media, depending on the options checked in the Decompose dialog, but non-timecoded material is decomposed just like all other kinds of video from tape.

Consider this: The sequence can always be relinked to the original media still on the drives, even after the decomposing. If the redigitizing of timecoded sources is complete and you relink the sequence, the sequence stays linked to the new, high-resolution, just redigitized media. The sequence looks to the media drives to try to fill the empty gaps left over from the missing non-timecoded shots. When the sequence tries to link, it is looking for timecode and tape name. It finds the original time-of-day clips and relinks them with the finished high-resolution sequence.

The goal of all these methods is to reduce the amount of redigitizing to a minimum. These strategies require selectively removing only what needs to be replaced. Many people are content to spend the extra money on drives and always work at high resolution. This is a perfectly fine way to work, although the computer may not react quite so quickly to commands that have to do with recalling and playing back all that high-resolution media. The computer handles the smaller, low-resolution files more quickly, and overall performance, like rendering, is better at the lower resolutions. As the drives and the computers get faster, this will be less of an issue, but for now it is something you might take into consideration as your project gets longer.

Re-creating Titles

If the titles created from the Title Tool and used in the final sequence were first created at the low resolution then they must be re-created at the finishing stage. This is an automatic process that takes little time and replaces the old titles with new, high-resolution versions. There are two parts to every title: the type information about drop shadows, borders, and alpha channels, and the title media created at a specific resolution when you leave the Title Tool. The type information must be used again when you re-create the title media, and new media is created based on what is set in the Compression Tool (on Xpress this is set in the Digitize window).

- Set the Compression Tool to the highest resolution possible.
- Mark inpoints and outpoints at the beginning and end of the sequence.
- Highlight all the video tracks.
- Choose Re-create Title Media from the Clip menu.

The Avid looks through the sequences, finds all of the titles, generates new media, and links it to the sequence.

In the process of re-creating titles, a warning says, "This process will remove any nested effects." This refers only to nested effects within the titles. Nesting video in a title is a very cool and easy-to-use effect. You fill a title with moving video by stepping into the title and editing new video over the unlocked video track (V2), and maybe you add some other effects inside. The problem when you are re-creating the titles is that the Avid is going back to the original title information. The original type information knows nothing about what you did to the title after it was created. In versions with a DSK, editing video into a title will make the title lose its uncompressed status.

You can preserve the work you did inside the title by subclipping the nested effects, dragging them to a bin, and then placing them back in the title after it has been re-created. The only way to subclip material from the inside of a nest, in later versions, is to use the Clipboard button.

- Mark-clip the entire material inside the nest.
- Hold down the Option or Alt key while clicking on the clipboard icon. This puts the material automatically into the Source monitor. Since the only time I use Clipboard is as Option/Alt-Clipboard, I have mapped the Option/Alt button—a little dot in the Command palette with the letters OP or AL—onto my Clipboard button.
- Use standard subclip techniques to drag the nested material to a bin.

In later versions, you have the option to use uncompressed titles and title rolls. You can't promote these uncompressed titles to 3-D, but you can play them in real time while there is another video track below with unrendered real-time video effects. Rolls are as easy as hitting the R button in the Title Tool and typing away. Crawls are one of the few things the ABVB does in real time that the Meridien does not. In the ABVB system, you will have a C button in the Title Tool for creating crawls.

Replacing Graphics

The final step for "rezzing up" is to bring all the graphics up to the finished quality resolution.

BATCH IMPORT

The most significant improvement to batch digitizing and re-creating a project from scratch is Batch Import. After you have finished digitizing the timecoded material, then it is necessary to reimport all the non-timecoded graphics, animation, and music (especially if it is from a CD). In the past, you would have to import all these elements at the high resolution and match them in by eye. Of course, the music does not have to be redigitized or reimported unless it has been deleted, since AIFF files will travel easily between Mac and NT at full quality.

You can bring all of your original material over to the Symphony through any of the methods we will discuss in the next chapter and then import them again using the Batch Import function. This reimport is frame accurate even without timecode.

Here is how to use Batch Import:

- Highlight the sequence in the bin.
- Choose Batch Import from the Clip pull-down menu.
- If all your graphics are offline, choose to replace all of them.
- Navigate the menus to the folder with all the graphics.
- They are imported and connected to sequence frame accurately.
- If your graphics are scattered across different folders, you will have to repeat the previous two steps.

There are actually two methods to Batch Import, depending on your desired result. The first method is to highlight a sequence and batch import all the files that are connected to that sequence. The system looks for the names of all the offline imported media and grabs those files from the folder you have pointed it to. Those files are linked only to that sequence. This is because you may have multiple sequences that are still at low resolution that you want to be left alone.

The second method also keeps in mind that you might not want to affect all the sequences on your system with the new high-resolution graphics. You go to the original bin that contained all the clips representing the offline imported

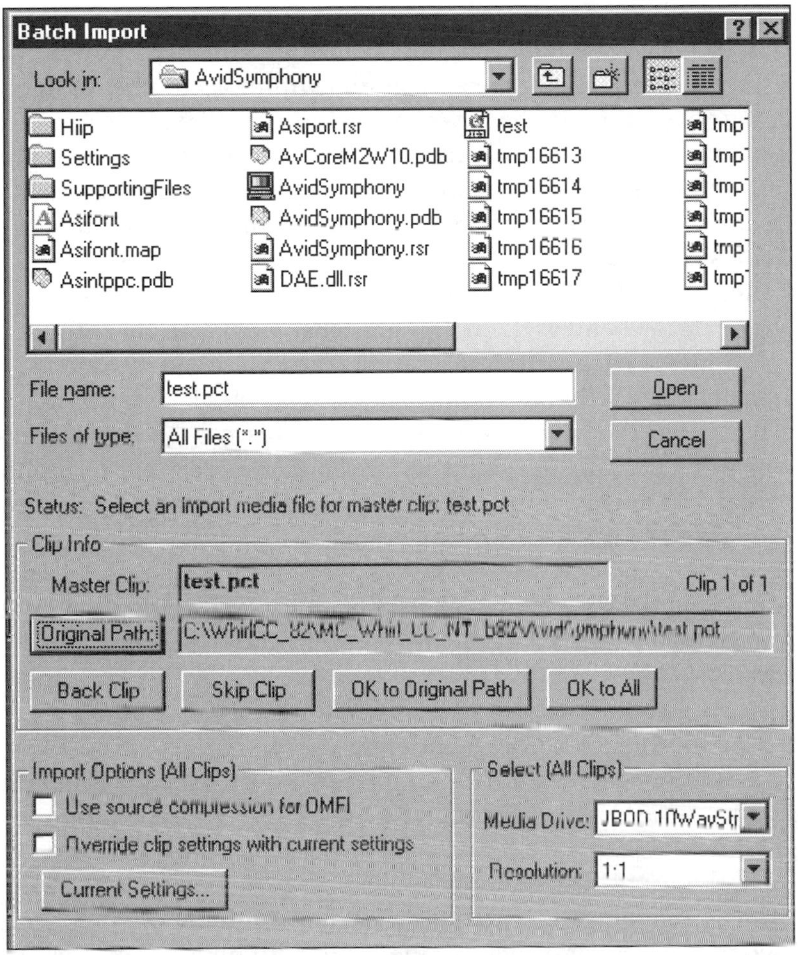

Figure 10.8 Batch Import Dialog

media. Select them all and batch import. You will now have a bin full of online high-resolution media, but this media is not really linked to any sequences yet. Now you must go to the sequence and drag it into the same bin as the new batch-imported clips. Select everything and choose "Relink to Selected" with the Relink command under the Clip menu. With the latest versions, there is a bit more wordy choice, which is "Relink all non-master clips to selected online items." Now that specific sequence has been linked to the high-resolution graphics. This method is best if you want to relink multiple, slightly different sequences all at once and they share many of the same graphics.

The hidden benefit to the Batch Import function is that identification of the graphic is based on file name. You can point the Batch Import file hierarchy to any file and it will import and link the new file to the sequence. This means that you can use it for versioning. If you have an updated graphic, just delete the media from the old one, navigate the Batch Import dialog to the new graphic, and batch import it. The system will link the new graphic frame accurately to the sequence in place of the old graphic. Perhaps you have a new version of the music used and all you need to do is import it and link it to the sequence where the old music was. The user can then change the name of the clip to reflect the version number. If you have been doing many nested layers or editing the music and cutting to the very specific frames on the beat, then this ability will save you hours.

For older, non-Meridien systems, this was not so easy. If your imported graphics have alpha channels, then they are considered "effects with source" and are treated differently than straight graphics. On these older systems, the only way to change the resolution on alpha channel graphics is to reimport them and cut them back into the sequence one by one. This can be done a little more quickly using the Replace function in Media Composer and Symphony. The Replace function is the blue arrow available under the Fast menu in the center of the Source/Record window or in the Command Palette. Use Replace instead of Overwrite. Replace works well because it requires no mark inpoints or out-points. In fact, any existing points should be deleted. Put the blue position bar close to the beginning of the new, high-resolution graphic in the Source window, and position the blue position bar in a similar place in the Timeline over the graphic. Choose Replace and the system will use the blue position bar as a sync point for both Source and Record windows.

Here is another method for fast replacement of material. This method is useful on older systems and edits directly from the bin to the Timeline.

- Use the Mark Clip button to select the low-resolution graphic in the Timeline. A quick way to do this is to highlight the graphic with the red segment arrow and then hit the Mark Clip button. You bypass having to turn tracks on and off.
- Choose the red segment mode arrow and drag the graphic from the bin to the Timeline.
- Hold down the Ctl/Cmd key, and the graphic snaps to the inpoints and outpoints in the Timeline.

If you created keyframed moves for these graphics (including fading up and down), you want to preserve that work for the new high-resolution replacements. To save the keyframes for a regular graphic:

- Go to the Effect Mode and highlight the graphic with the keyframes in the Timeline.
- Drag the effect icon in the upper left-hand corner of the Effect Mode Palette to the bin.

This saves the keyframes without the source material when there is no alpha channel. To save the keyframes from an imported graphic with an alpha channel:

- Option-drag the effect icon from the Effect Mode Palette to a bin.

AUTOMATICALLY RELINKING GRAPHICS WITHOUT ALPHA CHANNELS

Before Batch Import was available, I developed a way to automatically replace graphics or animations that do not have alpha channels. If there are only a few graphics to replace, the two methods just described are probably the best and the simplest. But if you have 20, 50, or 100 graphics, you need to do a little more work to set up a process that forces the sequence to break the links with the old graphics and relink to the new ones.

The trick here is that to properly relink to a sequence you need timecode and tape name to match between the source in the bin and the clip in the sequence. When you import a graphic, the timecode is always the same: 1:00:00:00. But what is the tape name? There isn't one, and that's the problem. You can give the imported graphic a tape name using the Modify function! If you give all the graphics the same tape name, that would be simple. Just select them all and modify them all at the same time. Unfortunately, this doesn't work. They would all look the same to the sequence with the same timecode and the same tape name. This would link all the graphics in the sequence to the first graphic modified—all the old graphics would be replaced by one new graphic! So you must give each graphic a unique tape name. Break the link between the old graphic and the sequence by forcing a relink of the old graphic master clip to the new high-resolution graphic media. If you used many PICT files or other graphics formats and combined them with effects, then this method saves you time compared to replacing them by hand.

- Determine the graphic files needed in the finished sequence and isolate them in a bin. This can be done by opening all the bins with graphics or the Media Tool, selecting the sequence, and using Find Media Relatives in the Sequence Bin pull-down menu.
- Highlight the graphics used and drag them into a new Used Graphics bin to isolate them.
- Click on the name of the first graphic (not the icon) and copy the name into the Macintosh clipboard (Command-C).
- Click on the icon of the graphic to highlight it.

- Go to Clip menu and choose Modify.
- In the Modify dialog, choose Source and paste the name of the clip as the new source. You are limited to 32 characters and no punctuation, but it's a start. The idea is to make the source name simple and obvious.
- Click OK on all the messages about how terrible this is and that you will probably die penniless, etc.
- Do this for all the graphics you need to relink.

Actually, modifying master clips is very serious, but in this case you need to modify a lot of individual master clips as fast as possible. You may consider programming a macro to just click OK on all those messages.

- Import the necessary graphics one more time at the higher resolution into another Reimported Graphics bin.
- Modify the sources of the new graphics. This time, when you go to Modify source, you see the choices for source names already in the Choose Other Tape Name window.
- Double-click on the obvious names that match the name of the graphic, and you have saved yourself all that cutting and pasting.

There is a new twist to this method in later versions of the software since you can specify the length of the graphic during the import process. You must import the low-resolution and the high-resolution graphic files at the same length. If you always leave it at the default length of ten seconds, you can't mess this up.

Now just delete all the original low-resolution graphic media in the Media Tool and relink the sequence. Make sure none of the boxes are checked in the Relink dialog. All your graphics in the sequence are relinked to the higher resolution media, and all of your effects that involved the imported graphics are intact (although unrendered).

The more complex procedure is to leave all the low-resolution graphics online and force the sequence to relink to the high-resolution graphics:

- Drag the sequence into the Reimported Graphics bin with just the high-resolution imported graphics.
- Select all.
- Relink the sequence with "Relink to Selected."

This allows you to keep all the low-resolution graphics on the drives linked to other versions of the sequence.

RENDER SETTINGS FOR MOTION EFFECTS

One of the hidden problems with going from offline to online using Avid systems is what to do with the motion effects. There are three kinds of motion effects (officially, anyway—perhaps you can find the hidden, unofficial fourth type in the later versions). They are duplicated, both, and interpolated. Duplicated fields is primarily used with single-field resolutions and is faster than the other two, so it is preferred by most offline editors. The problem for the online editor is how to change all of the offline motion effects from duplicated, which look just fine at AVR 4, to something that looks good at 2:1 or 1:1. This means changing all the motion effects to either both or interpolated.

Many people think that interpolated will always look the best, but this is not really true. Interpolated motion effects look best when the speed of the effect is something that does not divide easily by the 100 percent speed (30, 25, or 24 frames per second). Interpolation makes up for this mathematical difficulty by blending just the right fields, but it can do only so much. Interpolation will also potentially blend field one from one frame with another field one from another frame. Although this may result in smooth motion, it uses almost redundant information by combining the first fields. Half of the frame's information and resolution comes from the second field, which has been dropped out in the interest of smoothness. Therefore, both fields is actually the type of motion effect that will give you the sharpest playback since it neither drops out field two entirely (duplicated fields) nor combines multiple first fields (interpolated); however, both fields will sometimes give you jerky motion playback, so the choice is really up to the online editor as to which type looks better on the final high-resolution master sequence.

How do you change an entire, redigitized sequence from one type of effect to another? In the past, this was a rather obscure Console command. The user would type Motionfxrendertype, a space, and then a number from 0 to 2. The numbers apply to the motion effect type in the following order (in versions 2.x/7.x and later):

```
0 = duplicated fields
1 = interpolated fields
2 = both fields
```

Then when you render from in to out, all motion effects would be re-created using the desired type. If a particular motion effect still didn't look right, you could change the Console command number and render that effect "at position" to see if it looked any better. Remember that the best way to make any motion effect look better is to pick a speed that divides more easily into the 100 percent frame rate. Although this is definitely a last resort, it should be considered if you are close to that speed anyway and the change would not be noticeable in terms of the content of the shot.

Now, there is a new setting for motion effects in the Render Settings. It gives you the three choices, but it defaults to the setting that maintains any motion effect type that is already there. The offline editor may have been aware of the different types of motion effects and may even have been working in a low-quality, two-field resolution. Then you would not want to change them all to something else or you may use your "online eye" to improve them one at a time once the project is in your hands.

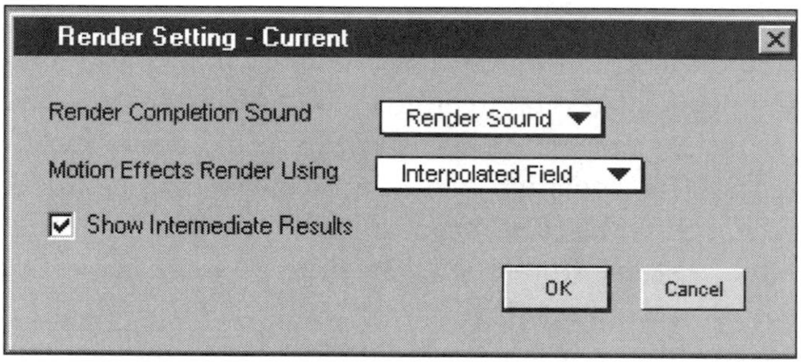

Figure10.9   Render Settings to Change the Motion Effects Type

While you are in the Render Settings, you should turn on the "Show Intermediate Results" checkbox and watch the client monitor the next time you render. It doesn't slow the system down at all, and it is amusing to the clients. Show Intermediate Results does happen to be useful if you see a part of an effect that needs to be tweaked. You can stop and tweak before you waste time rendering the entire effect. In some cases (where you haven't used acceleration or spline on the effect), you may be able to add an Add Edit before the area to be tweaked and then just render from there. This would take advantage of the Partial Render feature in the newest releases. To best use Partial Render, make sure the Render Range choice in the Timeline is set to partial.

After all the graphics and animation have been reimported and linked to the sequence, it is finally time to render. Most people don't realize that they only need to render the top track of any effect. As long as the top track covers the entire effect, you do not need to render all the tracks below. This saves a lot of time, but people insist on rendering everything anyway. It makes them more sure that they will not get a video underrun in the middle of a complex effect sequence and, if you have the time, this is foolproof. Rendering everything may also speed up access time when you are moving between layers because the system never has to render on the fly; it can just call up the rendered precompute. But if you are making many changes or are in a hurry to just get something out the door, then all this extra rendering can be skipped.

The new feature, ExpertRender™, was developed specifically for this situation. There is an in-depth explanation of the uses and benefits in Chapter 7.

SCRATCH REMOVAL

One of the most important stages to finishing any project that has originated on film is scratch removal or, as it is sometimes referred to, "schmutz busting." Perhaps this term comes from the ubiquitous Dust Buster and the fact that so much stuff that ends up on the final print is of an unidentifiable origin. Anyway, this process consists of many sharp-eyed people evaluating the final video image to catch any and every defect before it is signed off.

As the dirt, or "schmutz," passes by, it is flagged and fixed. With the new scratch removal tools available in the latest releases, this is a one-step process. By stopping on the problem frame, the user can click on the Scratch Removal button. This button must be mapped from the Command Palette and looks somewhat like an eraser. The Scratch Removal button will automatically put the user in the Effect Mode, put Add Edits one frame back and at the end of the problem frame, and put the Scratch Removal effect on this new segment. In one step, the problem area has been prepared for fixing.

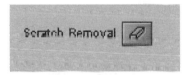

Figure 10.10 Scratch Removal Button

Where should you get the pixels to cover the scratch? You could grab them from somewhere else in the frame if there are many similar-looking surfaces. Or you could use the Paint Tools to clone a small section of the image and move it over the scratch. But most likely, you will want to take the good pixels from the last good frame. That is why the first Add Edit (those little white marks that indicate an artificial edit point) is placed one frame before the problem frame. Now the user has the ability to steal the pixels from the previous frame or the previous field (all effects in the Effect Mode are field based since 2.0.1/7.0.1). The default for this temporal cloning is automatically set to two fields, or one frame.

Generally, you will get a better looking fix if you take the pixels from a field one and apply it to another field one. This becomes less of an issue in the 24 fps progressive projects because there are no longer any fields, but it is still important in all video-based projects.

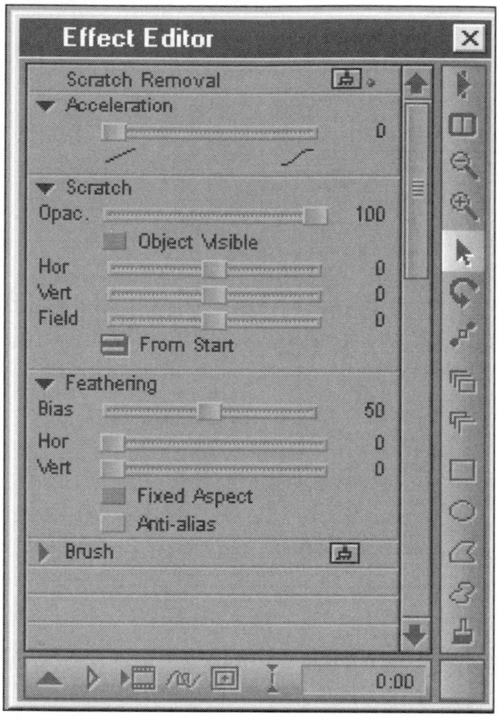

Figure 10.11 Scratch Removal Effect Interface

Sometimes the problem extends beyond a single video frame, as when the dirt is on a B or D frame in a film-based project that is being edited in 30 fps NTSC. Then you can trim the Add Edit to make the Scratch Removal effect as long as you like. The default setup is to take the good pixels From Start so the first good frame or field is used automatically. You can also set it to From End or Relative. Set to Relative, it will always change the field to maintain a relative offset. The temporal offset — the number of fields in either direction that can be used — is limited to ten fields.

However, you cannot use fields that extend beyond the limits of the Add Edits. This limits most fixes to only a few fields. If you need more than that, then you should bypass the automatic features of the Scratch Removal button and apply the Add Edits yourself. If you know that you will be adding a Scratch Removal effect to a long segment, then you should put your mark in and mark out points around the bad area. With your blue position bar set in between the marks, click on the Scratch Removal button. The system will put Add Edits at the marked points, plus one extra frame at the head of the effect for the reference frame. This last feature would be good for fixing the kind of thin vertical scratch that goes from the top of the frame to the bottom and lasts over a long period.

AUDIOSUITE PLUG-INS

With the latest release of the NT editing products (3.0/9.0/2.0), AudioSuite Plug-ins will be available and working for all models on all platforms. The special effects capabilities are amusing for all the times you need robot voices, but the real power of these plug-ins is in fixing everyday problems. They can be used either to lengthen or shorten an audio sound bite and keep the pitch the same, or to change the pitch and leave the length the same.

The best new capability using the AudioSuite Plug-ins is the ability to apply the effect to a master clip and create a new master clip. Thus, if you have a voiceover that is 31 seconds long, you can take the audio master clip and drag and drop it onto the AudioSuite interface. A new section of the interface will open up and show you the controls that apply to the master clip. After you make the basic effect decisions, you create a new master clip and can now use this as if it were a digitized source. From now on, every time you use the AudioSuite master clip, you can apply it into the sequence with the correction already done.

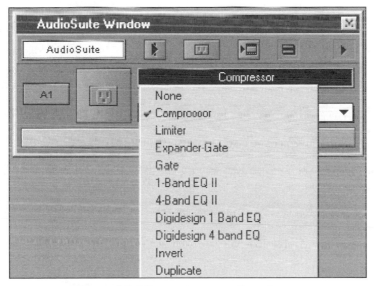

Figure 10.12 AudioSuite Effect Plug-ins

There are multiple benefits to applying an AudioSuite to a master clip. The first is that it allows you to apply an effect to multiple audio channels at the same time, or to a stereo pair. It also allows you to create a new master clip that is longer or shorter than the original clip. Instead of applying the effect in the Timeline and then trimming to accommodate the new length, you can cut this new clip in without any further adjusting. If you want to apply the effect to

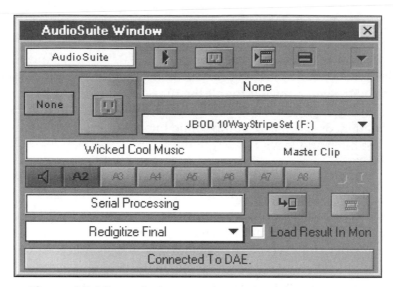

Figure 10.13 AudioSuite Interface in the Master Clip Mode

stereo pairs, be sure to check the Parallel Processing Mode. This will ensure that the effect is applied to both stereo channels simultaneously rather than one after the other (Serial Processing Mode).

CONNECTING A MIDI CONTROL SURFACE FOR MIXING

Ever since version 7.0/2.0, users have been able to connect an external Midi control surface to adjust the audio rubberbanding on the fly. This is a relatively easy connection to make, although it does take up one of the serial ports on the Macintosh and one of the com ports on the NT. There are switches for multiple devices to be connected to these ports if you really need two decks connected and the Midi control. You can minimize yanking cables with a flashlight in your mouth.

Follow the instructions in the manual for the connections for a JL Cooper Fadermaster Professional™. You are ready to start recording audio rubberbanding with all eight channels. When you are done recording, you can go back and filter out the unneeded audio keyframes to make it easier to adjust later with the mouse. To use the pull-down menu in the interface to filter the automation gain, you must first select the track(s) in the Audio Gain Automation window. In earlier versions, this was done by clicking on the Gang button, but in later versions, click on the button that shows the track name.

If you want to go back and adjust the levels after a first pass, you need to manually position all of the faders on the JL Cooper Fadermaster Pro. Both of the little lights next to the Track Name Selection button should be blue on the Audio

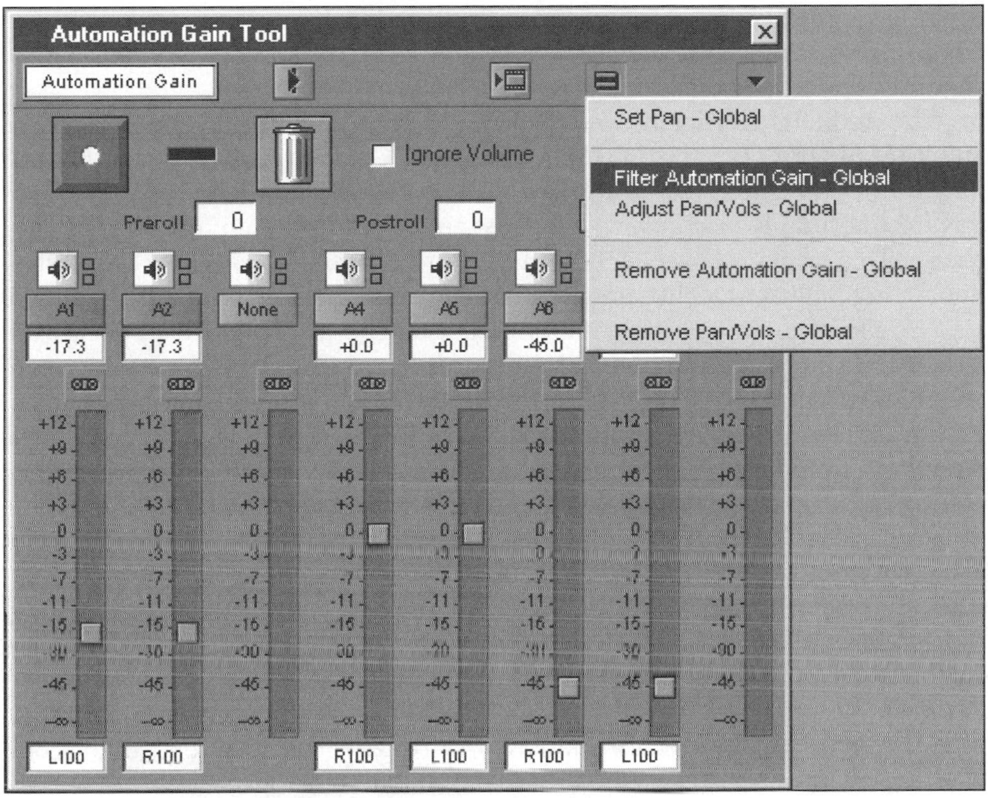

Figure 10.14 Audio Gain Automation Interface with Filtering Chosen

Gain Automation window. Now you know that you are starting with the faders in the proper place to begin recording, and pick up smoothly where you left off.

With the later versions of Macintosh 2.5/8.0 and NT 3.0/9.0/2.0), Avid officially supports the JL Cooper MCS 3000X and the Yamaha 01V, which both have flying faders. The flying faders are important because they will automatically reflect the rubberbanding level of any sequence as you begin to make changes. Start the audio gain automation recording and grab the faders as they are moving. The JL Cooper is touch sensitive, but the Yamaha is not. The Yamaha, however, can double as an audio mixer. You can use it to control all the ins and outs for audio sources and then switch over to become a MIDI control surface on demand.

IMPORTING AN EDL

After reading Chapter 8, you should be aware that an EDL is a terrible way to archive a sequence, however, there are times when you take a sequence to an online linear tape suite and assemble a project only to have it return to the Avid suite for more changes. Now you need to get the EDL into your Avid and digitize it. This is taking a non-compatible format and translating it to something the Avid can understand.

Using EDL Manager 1.9 or later, you can open an EDL and either save it as an OMFI file or import it directly to the Media Composer if they are both running at the same time. Use the arrow icon on EDL Manager that is going back toward the composer icon, and the list should import into an open bin of your choice. This brings in a sequence and a series of master clips that are almost 24 hours long. The master clips are there for reference and are not really to be used.

You may notice that these imported master clips do not have a project affiliation. This is a basic limitation of the import process. The OMFI files (which contain the project name) are created before the file is brought into the current project. You can select them all and modify the project name by changing the tape name (refer back to the graphics import section for the details). You will be getting new clips from decomposing the sequence, not by batch digitizing these unusable master clips.

- Load the sequence into the Record monitor.
- Use Remove Match Frame Edits from the Clip pull-down menu.
- Decompose the sequence and redigitize.

If your sequence does not import correctly, you must strip some of the complex information out of it.

- Open the EDL in Vantage (a word processor system in your Utilities Folder that ships with every Macintosh).
- Choose Show Invisible Characters in Vantage.
- Make all unusual effects into dissolves, strip out comments and GPIs, and basically make the list as simple as possible while maintaining all the original character placement.

Although EDL Manager 1.9 and later is more forgiving about missing spaces in the EDL, it is still best to leave everything in the original format. There are too many variables to guarantee that importing your list always comes in perfectly to the Avid, but you should get the majority of it without a problem if you simplify as much as possible.

RELINKING MEDIA TO AN EDL

If the digitized media from the original edit is still on the media drives, it is frustrating to not be able to link the imported EDL to the footage. The reason that this does not happen automatically is, as you will remember from the previous chapter, the importance of project name to relinking. When you import an EDL and turn it into an OMFI composition using EDL Manager, it then imports as a sequence into the editing program, however, there is no project associated with the source tapes in this new sequence. Without a project name connected to the tape names, there will never be relinking to media on the drives. Never, unless you have the newest versions of the NT software (3.0/9.0/2.0), although the Macintosh may not be far behind.

The most important aspect of the new relink choices is the ability to link media already on your drives to an imported EDL. In the past, this took many steps and was a rather unpredictable process. It was only made possible at all by following a workaround I published many years ago. It has now made it into a full-fledged feature.

Here is how to relink existing media to an imported EDL:

- Open the EDL in EDL Manager.
- Save as an OMFI file or, if the editing software is running, click on the arrow icon in EDL Manager that points toward the Avid editing software icon.
- EDL Manager will ask what bin to put the sequence if there are multiple bins open.
- Select the sequence in the bin and choose Relink.
- In Relink, choose Relink by Source Timecode and Tape Name and "Offline non-master clips to any online items."
- Make sure "Relink only to media from the current project" and "Match case when comparing tape names" is unchecked.

If there are multiple projects with the same tape name or there are multiple resolutions of the media, you may have to take another step. If you want more control over exactly what media is connected to this sequence, you should:

- Open the Media Tool and sift on the appropriate criteria for linking. This could be a combination of project name and resolution or multiple project names.
- Select all these sifted master clips and drag them into a bin with the imported sequence.
- Select everything in the bin and "Relink all non-master clips to selected online items."

There are still some obvious issues with relinking based on the text of tape names in an EDL. What if the tape names in the EDL do not match the names of the tapes assigned to the digitized media? There will be no relinking. This could happen because when the original EDL was made in EDL Manager, the type of EDL chosen could not support the original tape name. If you have a tape name that is longer than an EDL format can handle, then the tape name will be truncated. In other words, "Exterior Tape from Yesterday" becomes EXTERI, which will never relink. Since we are working from a simple text match, if the text is different, there is no automatic way to tell what it used to be. If this is the case, then you need to go through another step.

Here is how to force the link when the tape names have been changed:

- Check to see if any media has not relinked through the above process.
- Decompose the sequence with the checkbox set for "Decompose only those clips for which media is unavailable." This will create new master clips for all the areas of the sequence where the media is offline.
- Select the master clips and choose Modify from the Clip menu.
- Modify the tape name to be the same as the original digitized media that is already on the drives.
- If you have many clips from the same tape, you can modify them all at once. Otherwise, you need to do this to each master clip separately.
- Relink the sequence again.

This points out the importance of naming tapes so that they will fit into the restrictions of the most common EDL formats, even though you may have no intention of making EDLs at the beginning of the project.

Some people will use this feature to load sequences that have been created from a tape database. By using the database to create an EDL, they can quickly assemble sequences from stock footage, import the EDL to the Avid system, and relink.

The Old Method for Linking an Imported EDL to Existing Media

For those who have older versions of the software, here is the original workaround. I wrote this while Senior Trainer for Avid, and it is republished here with their kind permission. A warning: it is not simple, quick, or foolproof, but it does work most of the time.

In order for this process to work properly, it's important that you know the name of the project that your existing media is linked to and that the media is online. You can find the project name by looking at the Project column in your bins or in the Media Tool.

1. Make the "B reels" reflect the name of the original reel. You cannot link clips to media if the source names are not correct. The first problem is that some of the source names have been changed from the original project because they were turned into "B" reels for dissolves (dupes) in the linear suite. So reel 001 became 001B. After opening the EDL in EDL Manager, the system really doesn't know that the "B" is a B reel and will not make that assumption. Maybe there really is a reel 001B because it was the first reel you shot in Bosnia or Biloxi or Beantown. To remove

all the Bs, you need to open the list in Vantage. Carefully search and replace all the Bs with a blank space.

This assumes that the dupe material was created using the "multiple dupe" option. If you created your dupes using the "one dupe reel, new timecode" option, this method will not work for shots on your dupe reel because the timecode will not refer back to your original source material.

2. Import the EDL into EDL Manager version 1.7.3 or higher.

    Make sure the format of your list matches the format selected in EDL Manager. For example, if you are importing a PAL CMX 3600 list, make sure that EDL Manager is set to accept this type of list. Otherwise, you'll receive errors while trying to import. For version 1.9 of EDL Manager and later, you only need to indicate whether the list is NTSC or PAL. If you don't know the format of the list, guess until you're able to import without errors.

    Steps 3 and 4 are not necessary when working with EDL Manager versions 1.9 and greater. Use the arrow icon to move the EDL into the Avid system.

3. Save the EDL as an OMFI file.

4. Import the OMFI file into an empty bin. The OMFI file will import into your bin as a sequence and master clips.

5. Highlight the first master clip. Note that it will be almost 24 hours long.

6. Choose Modify from the Clip menu.

7. Choose Set Source from the Modify submenu.

8. Deselect "Show only this project's tapes."

9. Select the tape name that corresponds to the original project name and tape name. It will be in the format Project Name·Tape Name. You may see several projects associated with a tape name. If you don't select the correct one, the media will not link up. It's also possible you'll see more than one entry for a particular project and tape name combination. If this occurs, you'll have to try each choice in turn.

10. Highlight the sequence and choose Relink from the Clip menu. Do not select any of the options in the Relink dialog box. Click OK.

11. Repeat steps 5–10 for each master clip. The sequence should now be linked to your existing media.

12. Delete the new master clips [the 23:59:29(24) long ones created from the EDL]. The new master clips will never relink—they're used as intermediaries. You have actually relinked the sequence to the original master clips. You can now use Match Frame and Find Bin to refer back to your original source material.

13. Highlight the sequence and choose Decompose from the Clip menu. This is a way of checking if any part of the sequence didn't link. In the Decompose window, select the "Decompose only those clips for which media is unavailable" option. Master clips will be created for those parts of the sequence that didn't relink. You can now batch digitize

them. If no master clips are created, then your entire sequence relinked properly.

One possible reason that some decomposed shots may not appear is that the EDL was altered in online such that a shot was made longer than its original master clip. You will need to redigitize these changed shots to reflect the changes in the sequence.

FTFT

Another important method of relinking, FTFT (Film Tape Film Tape) relies on keycode. This method is extremely important when working with multiple film transfers of the same material. Assume that most film-originated productions transfer all of their dailies using a one-light or inexpensive film transfer designed to get as much footage transferred as quickly as possible. Then this one light is digitized at an offline resolution to save disk space and put as much footage on the system as possible at the creative stage of shot picking and offline editing.

At the end of this process, the quality of the image must somehow magically be improved and then output to the master tape. Just redigitizing the one-light transfer is usually not good enough. The magical bit happens first in the telecine where the telecine operator records the keycode information that is already on the film. The keycode goes into the file that will eventually be transformed into an Avid bin (using Avid Log Exchange as an intermediate, translation step). This keycode associated with a start and end of each master clip follows the footage throughout the project. Once it has been entered into the Avid project with film options or any other 24 fps native capabilities, a true negative cutlist (the list of shots used by a negative cutter) can be generated.

The editor or assistant, after the offline has been completed, creates a *pull list* using FilmScribe or another cutlist generating tool. This is a list of entire scenes to be retransferred with final color correction. The pull list is generated based on the keycode, and the telecine operator/colorist does the final transfer on only the selected scenes. This is all basic stuff so far, but we are just about to hit a snag.

The original one-light transfer was digitized into the offline system and used to edit based on the timecode of the original videotape. But the latest, greatest film transfer is on a new tape with different timecode from the original one-light videotape. How do you reconcile the two different timecode numbers that apply to the same shots? Video decks don't search and cue up based on keycode!

Relinking is a powerful stealth tool, so Avid has created a new capability: relinking by keycode. Go ahead and redigitize the new material at 1:1 quality based on the new files from the second color correction. Pay no attention to the timecode on the new tape. Both color correction sessions have used the same keycode, and they will be dutifully listed in the new bin.

Take the original offline sequence (or a duplicate just to be safe) and put it in the same bin as the new redigitized 1:1 material.

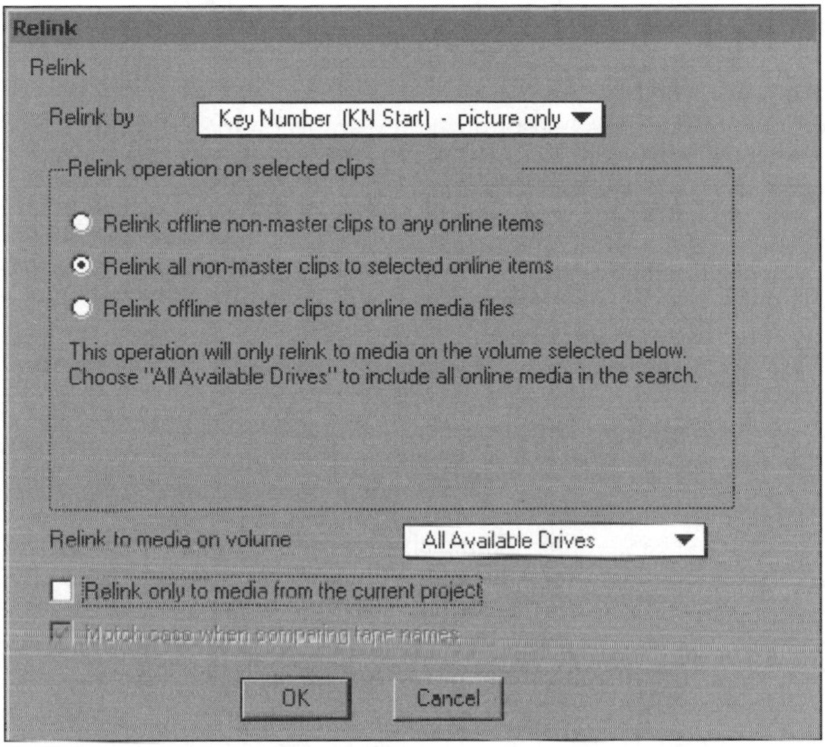

Figure 10.15 Relinking by Keycode across Multiple Film Transfers

- Select all the items in the bin (sequence and master clips).
- Choose relink from the Clip menu.
- Choose Relink by Keycode from the top pull-down menu in the Relink dialog.
- Choose the second radio button "Relink all non-master clips to selected online items."

For some reason, the majority of these second transfer bins are opened and used in a new project different from the original offline. This may be because the redigitized stage happened on another system different from the offline system and the digitizer did not have access to the original project. It doesn't matter, except that if you are redigitizing in a new project you must:

- Uncheck the setting for "Relink only to media from the current project."

The original offline sequence will link to the new, redigitized, retransferred material, and in an instant you have avoided the typical eye matching that must occur with other systems. Especially on longer, more complicated, or short-form multi-layered projects, this one procedure will save you hours of tedious work.

VTR EMULATION

There are times when the Avid system does not control a deck but is controlled by an edit controller. This is useful if you want to tie the Avid into an edit suite and use it like a disk recorder or another deck. It could hold bits and pieces that you will be using repeatedly for a linear online assembly. You can also use VTR Emulation to output the sequence time to tape and slave other devices to the Avid. This becomes especially useful when tied to an external MPEG compression system that can control a tape deck. You can compress MPEG directly from the Avid output when you connect it through VTR Emulation.

VTR Emulation requires a special cable that must be ordered from Avid, and then you can set it to behave like any type of VTR that works best with your system. VTR Emulation defaults to a Sony PVW-2800. The H-Phase and the SC Phase also need to be adjusted in the Output Tool just like any deck, so that they match the timing of the switcher. If you adjust the H-phase and the SC phase correctly, you don't get a giant horizontal shift or color change when you attempt to dissolve it with another source through a video switcher.

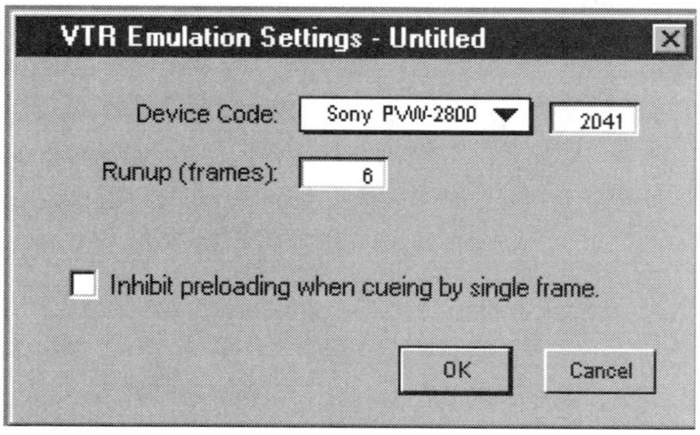

Figure 10.16   VTR Emulation Setting

CONCLUSION

The methods in this chapter are relatively advanced and complicated, but they are created in response to the unexpected needs of the real world. Even if you can't remember all the steps involved in these different techniques, at least you know that these procedures are possible. There is a lot of flexibility when it comes to going from offline to online. This extra power means that you must spend a bit more time and go through more steps when you are working on a complicated sequence. If you find yourself able to always work at the highest resolution, you can save yourself much of this effort, but that may not be realistic. When the time comes that drives are so inexpensive that most people can work at high resolutions all the time, then all of these techniques will be as obsolete as knowing how to thread a quad VTR.

11

Macintosh and Windows NT: Working Together

THE STATE OF NT AND AVID

There has been much gnashing of teeth, pulling of hair, and wasting of Internet bandwidth over which operating system is better: Macintosh or Windows NT. Although the Internet may have plenty of bandwidth to spare, we humans do not, so I would rather spend time discussing specific issues that apply to editors. Avid chose both operating systems and will continue to develop for both. The real choice comes down to whether you have a lot invested in Macintosh-based systems and do not want to switch or you need or want some of the new capabilities of Windows NT. The other question is whether or not you can make the Macintosh and the Windows NT systems live happily together. Some of the next section will deal with the Mac-to-NT workflow, and by the time this third edition is published, they will have been used by Avid for over a year and a half. This is hardly the bleeding edge, but it is a few extra steps. Once inside the editing applications, you will quickly forget the differences.

BENEFITS OF NT

The immediate benefit of Windows NT has nothing to do with the operating system, but with the freedom to configure a custom CPU with all the required processor cards. Since Apple abandoned its clone makers, the Windows world has the advantage of multiple six-slot, multi-processor, high-powered CPUs to choose from. This ability to work without an expansion chassis, as required on the Macintosh G3s and the latest G4s, means that there is no throughput slow down because of a PCI bridge chip. No amount of arguing about the aesthetics of an interface will change this throughput restriction and the problems associated with an expansion chassis. The G4 may change all of this with a faster PCI bridge chip, so this restriction may fall away along with other speed differences, even by the time this third edition is published.

There are some other benefits of NT, which editors will find right away. In general, the copying of files, even over a network, is blindingly fast. The operating system and Avid's implementation of Symphony, Media Composer, and Avid Xpress are very stable. Part of this stability comes from the NT ability to protect programs from each other, so when one program hangs or crashes, the others are not affected. Also, vital parts of the operating system are protected by restricted access through administrative privileges. This keeps the random freelancer from reconfiguring important files and causing the kind of intermittent problems that are so hard to track down over the phone. NT systems are favored by mission, critical facilities and server-based businesses because they will run dependably and with little administrator intervention.

My favorite NT benefit is the right mouse click, and Avid has taken advantage of this opportunity. Although there are programmable mice or track balls that allow the programming of a right button on a Macintosh, they are not application- or windows-specific like the NT operating system. When you right-click in the Timeline, you get Timeline-specific shortcut choices, and this type of context-sensitive list is available throughout the interface. As you begin to use the right mouse more often, you will find yourself searching less for often-used functions and using the mouse less for sliding across two monitors for a pull-down menu. In the Macintosh versions 2.5 /8.0, there is some of this context sensitivity. The user must hold down the Shift-Control keys and click in a blank space in the interface, but somehow it isn't as simple or convenient and doesn't get used as much (yet).

NT allocates as much RAM as an application needs, and the Avid editing systems need a lot of RAM. There is no more confusion with a call to Support because someone changed the RAM allocation of the software or the system's RAM cache while you weren't looking. The way that NT uses virtual memory is also very efficient. Virtual memory is the process where the OS uses a hard drive as an extension of the RAM. When the RAM fills up, the OS decides what files can be stashed away on a protected part of the internal hard drive. Avid configures a 300 MB file just for this so-called "page swapping." Although other operating systems do this, the NT system does it in 4 k chunks. Other systems may have a 64 k section that is used for page swapping. If the file to be swapped to hard disk is only 3 k, then 61 k is wasted when the virtual memory swaps the files from RAM to disk and back again. This means you can operate more applications at the same time, and the performance of the system is still very fast.

There is preemptive multi-tasking and multi-threading on NT, and some applications can take advantage of multi-processor rendering or network rendering. Although this will be possible with Apple's OS X, it is possible today on NT and seems to have all the bugs shaken out. Avid editing programs do not currently take advantage of some of these NT benefits because Avid relies on hardware that cannot be shared during very critical operations. Even the programs that allow you to multi-task will not allow those tasks to use the same piece of hardware at the same time (like a video board). Sure, you can do a few things while rendering, but most important tasks with a video program do require access to the video board. If a system allows you to play video and render at the same time, then it is not using the same hardware for both. Avid uses the video board to accelerate most rendering, so although you do not share the video board, there are already some everyday benefits.

Other graphics and compositing programs will allow you to take advantage of the multi-processors, and the Avid systems are now multi-processor safe. Some AVX plug-ins, like Artel's Boris Effects, can use the multi-processors when used within the Avid system. So you could set up a render in the Xpress, click on the toolbar and go off to work in another program, like Adobe Photoshop, while the Avid rendering continues in the background.

The Task Manager is my second favorite part of NT (perhaps because I work with so much beta software). Control-Alt-Delete will bring up a dialog that allows me to look at all the programs being run and force one to quit while all the others continue on. This flexibility and protection from bringing down the whole system means I can get all sorts of work done while copying, downloading, and rendering without fear of interrupting an important background process.

I have three favorite NT shortcut keystrokes. The first two use the Windows or Start key and operate no matter what program you are using. The first is Start-M, which minimizes all the programs you are using. This clears the desktop and allows you to jump into some new thing in a single step. The other keystroke is Start-F, which allows you to go into a system wide search in one step. Control-F generally lets you search within the program you are using, but Start-F will search everything on your drives or anything on the network. You can use Start-F to search for a file that a graphic artist has placed on their local drive if you have set up your ethernet or Unity network that way. The last keystroke is Alt-Tab, which allows you to cycle through all the programs you have running. If you are jumping between an Avid editing program and a graphics program, this will save you many mouse clicks.

DRAWBACKS OF NT

Some drawbacks to this operating system apply to editors as well. This is to be expected with any general purpose operating system that is forced to work in a very demanding, specialized application. The most obvious difference is the lack of a third modifier key. Because one of the modifier keys, the Windows or

Start key, cannot be used by Avid (or anyone but Microsoft), certain power user keyboard combinations cannot be reproduced from the Macintosh. Most of these are now stand-alone buttons and put in the Command Palette. These functions can be mapped to the keyboard for single-button use. Now that these power user functions exist as buttons, they are actually a bit easier to use, but you must have the foresight to map them to your User Settings before you begin to work. In many cases, the Shift key has been brought into play to add an extra level of functionality. The F1 key is also reclaimed by Microsoft to forever call up the Help menu.

The main drawback for Macintosh users who switch over to NT is the switching of the Command key with the Control key. These keys serve similar purposes in the two operating systems but are placed differently on the keyboard. The Control key is on the outside of the keyboard on NT, whereas the Command key for Macintosh is positioned next to the space bar. Making your fingers remember which system you are using is quite a challenge.

Windows NT does not recognize Macintosh disks right out of the box. This is why Avid recommends (and includes with some systems) a third-party program like MacOpener™ from DataViz. Once installed, the only drive compatibility problem is with Macintosh striped drives. NT striped drives can move between NT systems because the Avid Disk Mounter creates a special file on the striped set. We will go into some detail about this topic later in Chapter 13.

Lack of privileges can sometimes keep a user from performing some important troubleshooting procedures or from connecting to a network. If you need everyone to have access to everything all the time, then make sure that everyone has Administrator privileges. If you are a bit more cautious about the openness of your mission-critical editing systems, then you may want to give those privileges to only a few authorized people. Those people should be on call to react to problems that non-Administrators cannot or should not handle whenever they arise. After your password has been changed a few times by others who think that's pretty funny, you may start to limit privileges.

The last NT problem is one that should change with the release of Windows 2000. Because Avid requires two monitors to work like one large monitor with different bit depths and video capabilities, they have made their own Edit Display Controller card (EDC). This card has enough RAM and resolution for the specific Avid RGB monitor needs, including being able to play video at full frame rate. Unfortunately, this means that all error or operating system messages come up in the space between the two separate monitors—what NT considers to be the middle of the screen. In order to read them, you need to click on the messages with the mouse and drag them to the middle of one of the monitors. Of course, if you don't need to read all the text of the message, you can just click on OK or hit the return button and be on your way.

EVERYDAY EDITOR ISSUES WITH WINDOWS NT

Administrator Privileges

If you never want to bother with permissions or user IDs or different passwords, then you can make everybody an Administrator. You may find, with many users, that this will lead to more downtime in the long run. If everyone can install anything onto your editing workstation or muck about with important system files, you will eventually find something that conflicts with the Avid software and specialized hardware. You must weigh this possibility against the potential frustrations of a user who needs to do something important and time critical when there is no Administrator around.

In general, one person should be responsible for the upkeep of the system software, and everyone else should get his or her own password and user ID. Although this may take some getting used to, overall, the limited access will keep billable systems in a billable state. The Administrator can call up records of who used the system and when, which can help for billing. This may also allow better troubleshooting since the last person to use a system can answer many of the basic Support-type questions, and all of their actions are automatically recorded in the Event Viewer (under Administrative Tools).

Fonts between NT and Macintosh

Be sure that you have the exact same fonts on both the Symphony system and the Macintosh offline system and that you are using versions 2.1/7.1 or later on the Mac. This is because NT and Mac fonts do not play well together. You may find that you need to actually take all the fonts you have in one format and convert them to another format (Mac to NT and vice versa). There are some shareware programs like CrossFont from Acute Systems (www.asy.com) and, of course, the professional shrink-wrapped application Fontographer by Macromedia if you need to convert lots of fonts. If you find that the correct font is not on your NT system, then you will be prompted during the Re-create Title Media function to pick another font. Be aware that this font, or even the same named font on NT, may have different kerning, leading, and point size. You should probably check all of your titles after re-creating a project on an NT system if you began on a Mac. Once replaced, the new font will be used on every title that originally used the missing font.

If you have Adobe Type Manager® on your Mac system, then you will not see the same level of font manipulation (smoothing, kerning, etc.) unless you have it on the NT system, too. The font replacement information can be copied and moved to another system or deleted since it is only a Site Setting. It is called AvidFontSub.avt and is located generally on the C drive in the Settings folder. It is created the first time you substitute a font.

If you have used any version earlier than 2.1/7.1, then you will need to open the project in the latest version of the software. Once opened in the later version, you will need to open every title and save it again. Clearly, this could be tedious, so make a special effort to offline your project with the correct, supported Symphony-compatible version, 2.1/7.1. If you do not have suitable fonts that match across platforms, then, as a last resort, you can always export a title from the Mac as a PICT with alpha and import it to Symphony as a Downstream key.

Since all system owners who purchased 2.0/7.0 should have gotten 2.1/7.1 as a free upgrade, make sure you get your copy. You may find that an earlier version of the Mac software may come across just fine, depending on the complexity of the project. But you will have to confirm everything by eye once the project has been redigitized to make sure that absolutely everything came across without problems.

Using Quicktime

There is a more in-depth explanation of Quicktime issues in Chapter 6, but here are some basics. The first releases of Avid Xpress 2.0, Media Composer 8.0, and Symphony 1.0 on NT did not support Quicktime import or export. The user needed to use a Quicktime-to-OMF converter (that shipped with every system). You would use this QT-to-OMFI converter on a Macintosh system to change the Quicktime movie to OMFI. Then you would import the OMFI file into the NT.

A Quicktime Codec can be used on NT for creating Quicktime files on third-party programs for import to the Macintosh Avid systems. This Codec was made available with the Macintosh Media Composer 8.0 release and is identifiable by the ability to save Quicktime with the new Meridien board video resolutions (Codec version 8.0 and later). The most confusing problem is that during the majority of the development of this first NT-based editing system (Xpress 2.0 and Symphony 1.0), there was no cross-platform Quicktime version for NT available. Toward the end of the development, Quicktime 3.0 hit the market and Avid began to make the next version completely compatible. When the latest NT editing systems came out, they were completely compatible with Quicktime 4.0. Confusing? Yes, this is an example of two developers missing schedules by just enough to perpetuate a problem for almost ten months between major releases.

Now there is a Quicktime Codec for both Macintosh and NT that will make cross-platform movies that can be imported to either system (Codec version 9.0 and later). Be sure to check the cross-platform choice when making the movie on a Mac if you plan to move it over to an NT system in the future. Otherwise, you will need to open the Macintosh-only Quicktime movie in a program like Media Cleaner Pro by Terran Interactive and save the movie as cross-platform.

The newest versions of Avid Codecs and editing systems will recognize an alpha channel embedded into a Quicktime movie, a 32-bit file. If the system cannot recognize a 32-bit Quicktime file made with the Avid Codec (version 7.2 until

2.5/8.0 on Mac, and 3.0/9.0/2.0 on NT), then any movie with an alpha channel made with the Avid Codec will be a *slow* import. Slow import also means that you are changing the resolution of the file during the import. This has implications if your alpha channel is a different resolution than your fill. If you know you are going to an older system for import, you are better off using the earlier methods. Create two separate files, one for the foreground and one for the alpha channel (see Chapter 6 for more details). Consider if you are going to use the same Quicktime movie on multiple systems. If some of them *can* take advantage of the new 32-bit Codec, then you may want to put up with the slow import on some systems just to enjoy the simplicity of moving around a single file.

So before making any Quicktime file, it is best to know where it is going to be used. If you don't know, then you *must* make it a cross-platform movie. You may also not be able to take advantage of embedding the alpha channel using the new 32-bit Quicktime Codec. You must be willing to accept a slow import at some stage of a project if you use an earlier version of the editing software.

With the releases of 3.0/9.0/2.0 on NT, many of these problems have been eliminated and in fact will allow a 32-bit Quicktime file to be imported to NT. This is an excellent reason to upgrade your NT editing systems and make sure that all of your third-party graphics and animation experts have loaded the new Codecs on either the Macs or NT systems that they are using today. Avid has made a concerted effort to place all the most current Codecs onto their customer website (www.avid.com) under the Technical Support section. You may find it easier to regularly point your colleagues to the website to download the latest and greatest versions to stay current with what you are using.

In summary, if you are using the latest versions of editing software on either platform, you can import and export Quicktime movies using the Avid Codec and embed an alpha channel. If you are using the first NT releases, then you cannot import or export Quicktime and must use a Quicktime-to-OMFI converter on a Mac system before sending the OMFI to the NT system. You should make sure all vendors and all workstations used for graphics and animations have the latest Quicktime Codecs for either Mac or NT and check the Avid website periodically for any improvements. Finally, if you are working on an up-to-date system and must move material backward to an older system, you may want to make your Quicktime movies without an embedded alpha channel when using the Avid Codec. Include the alpha channel as a separate movie and combine it with the foreground as a Matte key while in the editing system.

CREATING ANIMATIONS

The main strategy for rendering animations in third-party programs for use on both the offline and finishing systems requires rendering the animation twice using the Avid Codec. You would render it once as the offline resolution and once as uncompressed. This is not that much extra time if you are rendering overnight or you are continuing to render multiple draft versions at lower resolution until you get it just right. After the animation has been approved on the Mac, you can render the second version for the NT at the higher resolution. This will give you the faster import on the Mac, and when it comes time to import the animation to Symphony, you can use the Batch Import feature to have it automatically go into the right place in the sequence. Just point the Batch Import dialog to the new file and it will replace it as if it were the original file.

MOVING FROM OFFLINE MAC TO ONLINE NT

Moving from offline to online on a different system with different video hardware limits what you actually want to bring across. Mostly you should be interested in three things: the project (including bins, sequences, and original graphics), User Settings, and the audio. To be totally compatible with sending project information to the Symphony and back again to the Mac offline, you should be sure to use version 2.1/7.1 or later. This is the earliest version that is fully compatible with roundtripping. There is also a new setting in the General Settings of 2.1/7.1 that keeps you from naming a project, bin, or any clip or sequence with characters that are incompatible with Windows.

General Settings (Current)

☐ Drive Filtering Based on Resolution

Default Starting Timecode   `01:00:00:00`

Default Starting Edgecode   `001-00000&00`

☒ NTSC Has Setup

Project Format

These settings are a reflection of the current project. To ensure all new projects are similar, drag this setting to your 'Site' setting.

NTSC
16mm
Film Style Editing

Audio File Format: OMF(SDII) ▼

Cancel

OK

☒ Use file names compatible with Windows®

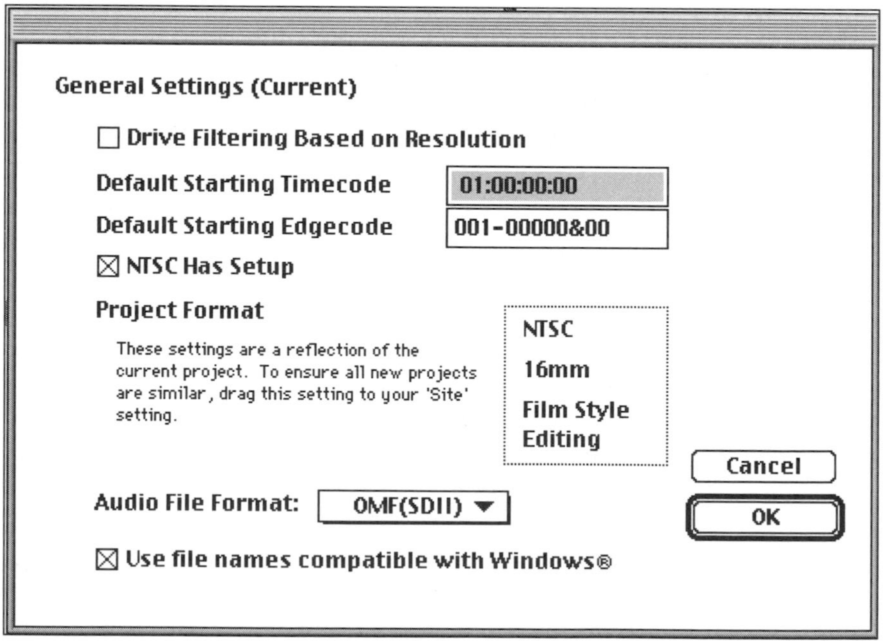

Figure 11.1   Windows-Compatible Setting in the Macintosh General Setting

SENDING METADATA

Technically speaking, bins, projects, and settings are data about data: metadata. They are the most important part of any project that needs to be re-created. Although, at the very least, you need only the sequence bin, having all the bins, just in case, is an excellent idea. The only question now is: "What is the best way to send that data to the Symphony system?" There are two major choices: removable drive or a network. Which one is best for you depends on the size of the files.

First, you must decide what else you need to send besides the project and bins. In all projects, after 2.0/7.0, there is a file called Statistics. This is a folder full of usage records as text files. In 2.1/7.1, there is a limit of 25 Statistics files saved before they are replaced by new ones, and the Statistics file is not updated as often, which tends to speed up performance. If you have been using an earlier version of 7.x, you may have many more Statistics files than 25, so pick out just the most recent. Consider that if you bring the Statistics file over to Symphony, it will continue to increment the usage amounts correctly. Decide if you need to take up the extra space and time it requires to bring them over.

In 2.1/7.1, the Attic now saves a separate set of files for each individual project, so you should think about bringing that over as well. Have you been good about making multiple copies of every sequence at crucial points in the edit? If not, then you may have to fall back on the Attic as a last resort when the producer claims, "It was better on Thursday. Can we see that version?" And then hope that you still have the Attic that far back (there is a built-in bias to hold onto the previous day's Attic files in 7.x/2.x and later). If you want to copy the Attic for your project, make sure that you put the folder inside the new Attic folder on the Symphony system.

There will come a time when you will need to compress the metadata to make it fit on a floppy or send a large file over the network. The most common way to compress data on the Mac for the Wintel platform at this time is Mac WinZip. *Do not use this program!* It will corrupt your bins and you will have little success opening them. Instead use Dropstuff shareware from Aladdin Systems that comes with Mac OS 8.x. Stuffit Expander 1.0 for the NT is also shareware from Aladdin and can be used to uncompress when the data arrives on the NT platform (www.aladdinsys.com). These utilities have been licensed by Avid and thoughtfully provided with every Symphony.

NETWORK

Let's take a look at setting up a network between Windows and Macintosh. Project and bin information is really not that large. A long format sequence bin with several sequences could be ten megabytes or more, but compared to the size of the media, this is insignificant. So you really could use 10 Base-T Ethernet and AvidNet to send this information. If you are concerned with sending large amounts of graphic and sound files, however, you should consider only 100 Base-T or faster.

The best way to set up a multi-station network between these two competing operating systems is through a server. If you use a Macintosh as the server, you will need to use a third-party application on the Macintosh server, like Dave 2.1 by Thursby Software (www.thursby.com). If Symphony becomes your "server" as well as your finishing station, then you can use PC MacLAN by Miramar Systems (www.miramarsys.com), but use only version 4.1. Version 4.0 will cause NT service pack 4 to crash horribly and force you into the ancient and confusing world of DOS patches. These are pretty standard applications for connecting NT and Mac, but they are technically not supported by Avid. This means that the full range of usual testing and documentation has not been completed. If you choose to use them, you must figure out the details yourself using the directions and the manuals provided by the manufacturers.

You can also use an NT server between the Mac and the NT system and use the NT server package for Macintosh. The NT server operating system is more expensive than the standard NT workstation, and be prepared to learn a lot about domains and administrative privileges if you are a Macintosh user. Again, Avid does not completely support this configuration, although it is quite standard for other common connection needs.

The simplest answer is to just connect the Ethernet cable between the two systems and use AvidNet. AvidNet is Avid's Ethernet transfer protocol and network software and allows users to just drop information into the "in box," send it over the Ethernet connection, and open it in the "out box" on the other side.

If you have no existing network already, this simple Peer-to-Peer connection may be the answer for you. Consider a simple US$50 five-port hub between the systems. If you want to expand this network to handle multiple systems, however, you may want to revisit the server solutions. As mentioned in an earlier chapter, the good news is that although AvidNet serves as an easy Mac-to-NT-to-Mac translator, it is also usable no matter how complex your network is.

There are certain limitations for AvidNet transferring between Mac and NT, but most of the common needs are handled well. Check the matrix of what is supported in the AvidNet Release Notes for the details. If you are going between a Mac system with local Mac drives and an NT system with local NTFS drives, then you should be just fine. Make sure that the clocks on every system on the network have the same time or you may end up replacing a new sequence version with an old one based on a wrong time stamp. Keep AvidNet in mind as the easiest way to get data between your Avid systems.

REMOVABLE DRIVES AND SNEAKER NET

Unless you are clever about setting up Wide Area Networks (WANs) over the Internet, you may still need to rely on generic removable formats like floppy drives, Zip and Jaz drives, and MediaDocs. Perhaps the chat room–surfing public can get along without floppies, but professionals who still exchange data the old-fashioned way will still be hooking floppy drives up to our Macs. They are especially important considering that the most common use for floppies is to get logged information into the system via AvidLogExchange (ALE), FLEx files, or making EDLs. A single floppy can hold hundreds of ALE files.

Unfortunately, when you launch NT right out of the box, you will find that it is clueless about reading drives formatted on its operating system rival. There are several shareware and shrink-wrapped choices to easily rectify this apparent oversight by Microsoft. The first is MacOpener 4.0 by DataViz (www.dataviz.com), which ships with every Symphony system and is officially approved by Avid. After installing MacOpener, the NT operating system will recognize Mac-formatted drives and floppies and attempt to recognize Avid bins and projects. An Avid bin should have an ".avb" extension, and an Avid project should have an ".avp" extension at the end of the file name. All Avid bins and projects coming from the Mac versions 2.1/7.1 and later will be handled correctly. An NT system can only recognize what kind of a file it is dealing with by the extension at the end of the file name. If you ever have an Avid bin that has come across to your NT system as a generic file icon, you can add the ".avb" extension to the end of the file name and automatically the NT system will recognize it and treat it correctly as an Avid bin. You can always open a bin from inside the application, so even if the extension is missing, you can still use it.

There is one area where even MacOpener may not help you when going from Mac to NT, and that is when you use the dreaded illegal characters. Never start or use these characters anywhere in a bin name: _, /, \, :, *, ?,", <, >, or |. You will be given a truncated, modified file name. For obvious reasons this is as undesirable as it is easy to do. If you find that you have done this by accident or on an old project, there are shareware renaming applications like SigSoftware Name Cleaner 1.8.1 (www.sigsoftware.com). Again, 2.1/7.1 has a new checkbox in the General Settings (defaulted to off) for maintaining Windows legal names. It only takes a few attempts to open a bin called "Scenes 22/23/24" on NT before you realize the value of this precaution.

The other choice is to format all removable media on the Macintosh with a DOS format. You can do this easily using PC Exchange 2.0.2, a Macintosh extension that ships with all recent Macs (7.5 or higher). You need the version that ships with Mac OS 8.1 or later to be fully compatible with Windows file names. PC Exchange 2.2 can read names longer than the 8.3 format of Windows, up to 31 characters. If you do not have the option to format a floppy with a DOS format or you get a message that your DOS disk is unrecognizable, then you probably do not have PC Exchange enabled. Move PC Exchange to the correct Extensions folder in your System folder and restart the Mac. If it has been thrown away in an earlier attempt to clean out your system to the bare minimum, then it must be reloaded from the Mac OS installation disk or borrowed from another Mac.

MOVING MEDIA

If you are moving from offline to online, why would you want to move media? What needs to come over from an offline project that will not change in quality? Digital audio. Since audio is not digitized at an "offline quality," you would want to copy your audio media over to the Symphony system. You do not want to play from a Mac-formatted media drive on an NT system, even with MacOpener, so the first step after connecting the drive is to copy audio to the Symphony audio drive. Since the Symphony system uses SCSI for the audio, make sure you copy all audio files (and render all audio effects) to the SCSI audio drive and not the Fibre Channel JBOD (the drive tower) while working on Symphony.

In order to move audio easily, you must first digitize it on the Mac as an AIFF file, not an SDII file. This setting can be found in the Audio Settings of versions 2.x/7.x and later. This Audio Setting must be correct before you begin to digitize the offline project. If you plan to bring a project from an offline system directly to Symphony, then you should always digitize your audio as AIFF. If you are going first to a ProTools system or another digital audio workstation (DAW) that uses SDII files natively, then you should stick with SDII. The newest versions of ProTools do handle AIFF files, so ask the audio post facility what they want. The audio professionals will have to send you back the finished version as OMFI or on digital audiotape. If you make a mistake and digitize as SDII when you really want to go to Symphony, then you can export your audio as OMFI with media. By exporting as OMFI, you are secretly converting the SDII audio files to AIFF.

The only wrinkle in bringing audio media to Symphony is the lack of support for AudioSuite plug-ins in the first version. The media will be playable, but the effect in the Timeline will be unknown, and if you trim it, changing the length of the effect, you will unlink from the audio media and not be able to render more. To keep this unlinking from happening, you may want to take the AudioSuite plug-in on the Mac and make it into an audio mixdown. This will turn it into a standard audio master clip that cannot be unlinked from its media while trimming. All the recent NT editing systems now support AudioSuite plug-ins, so this is no longer a consideration (but ask what version they are using anyway).

There are some reasons to move video, but not many. The Meridien board does just about everything better with signal quality than the ABVB. The only real reason to bring video from an older offline system would be because you have lost the original tape or you have spent a long time color correcting something at

AVR 77. Opening AVR 77 as 2:1 doesn't make it look better, but it shouldn't make it worse. If you want to move video, you can export it as an OMFI file and import it to the Symphony system. Or you can just copy the files and import as OMFI (the media files are already OMFI native). You must change the Import Setting on the Symphony (or any Meridien system) to change the field ordering to odd. This is important because the field ordering of the video frames is different between the two boards. Meridien starts displaying the odd field (field 1) first, which follows SMPTE standards.

Since the compression schemes between the two boards are completely different, this will be a slow transfer because all the video will need to be uncompressed and then recompressed on the Symphony system. You may find that it is faster to redigitize in many cases if you have the source tapes. If not, beware of going past the two-gigabyte file size limit for any file on the Macintosh when you export as OMFI composition with media. You can perhaps export your sequence in smaller chunks. Remember also that an OMFI export with media takes the entire original file. If you want to take only the media needed for the sequence, then you should consolidate the sequence first to create smaller media files.

So to move video from the Macintosh offline system to the NT Symphony, you should consolidate the sequence and export as an OMFI file with media. This may be your only choice when you are working with non-timecoded video sources, material restored from a DLT, or material for which you have lost access to the original sources. Of course, you can always just digitize the digital cut from the offline session for scratch video.

IMPORTING AUDIO

You must still copy the selection from the CD to another drive before you import into a Mac or an NT editing system. A range of shareware applications called CD rippers can be downloaded from the Internet, or you can use the application that comes with the Symphony system called WinDac. Use WinDac to copy the CD selection to the NT desktop as a ".wav" file or to a media drive and then import to Symphony. If you have used the CD audio during the offline on the Mac and you are not bringing all your mixed audio files over for the finishing stage, then you can use the Batch Import process like with graphics.

ROUNDTRIPPING

Roundtripping is the term Avid uses for bringing a project back from Symphony to an offline system on the Mac. This should work well for metadata (2.1/7.1 or later only), but bringing video media is not supported; however, if the original media is still on the offline systems, then bringing back the clips and sequences from the NT system should link right up. This would be useful if you wanted to make a finished high-quality version, making changes on the Symphony and then bringing that new version back to the offline system to continue work. With time on the Symphony system probably more expensive than on the offline systems, this could save you some money on multi-part projects with long post-production schedules. Again, roundtripping is only supported with version 2.1/7.1 or later.

DISABLING THE DOWNSTREAM KEY

The ABVB systems have the ability to keyframe a move of a graphic in the Downstream key (DSK). This ability has been lost with the change over to the Meridien hardware, so you must take an extra step when moving a sequence with keyframed DSK. If you do not disable the DSK, then all of the keyframes will be ignored. If you disable the DSK, then all the DSK effects with keyframes will be changed over to regular effects. The only real implication of this is that the DSK effects are considered uncompressed quality. If you are moving the project to a 1:1 resolution on Symphony, then the quality difference will not be noticeable, but you will lose the potential of having the graphic real time with another real-time effect on the track below. ExpertRender will figure this out for you, and you may have to render a few more effects than if you had stayed on ABVB.

After you load the sequence into the Record monitor on the Meridien system, go to the Console. Type "MeridienDSK TRUE" and hit Return. If you want to reverse this procedure, type "MeridienDSK FALSE."

CONCLUSION

There are lots of industry standard answers for getting material between Macs and NT systems. Networks are clean and simple once set up, but unless you are on Unity, they may be too slow for moving video media. With the procedures and network connections in place, the main workflow is to go from offline Mac to finishing on NT with Symphony.

12

Finishing On Symphony

Avid Symphony and Symphony Universal are Avid's top of the line finishing systems. As such, they receive Avid's biggest, coolest innovations for the high-end editorial space. They are designed specifically for the kind of high-end projects that demand the best quality and the most streamlined workflow. When you finish on either of these Symphony models, you are getting the best conform from an Avid offline with features designed specifically for the fast-paced world of the front room edit suite.

This chapter will go into detail about the benefits of Avid's Symphony as it relates to the finishing of offline projects begun on the Media Composer or Avid Xpress.

DESIGNED FOR FINISHING

The Symphony system is designed for high-end editorial programs that require redigitizing from the Media Composer or Avid Xpress offline. At the finishing stage, all the pieces are assembled from the production process. Audio is laid back and graphics are inserted at their highest quality and final version. Color correction is applied shot-to-shot while blemishes and mistakes are covered up or corrected with scratch removal. Of course, you can still edit during the finishing stage—one of the real hidden benefits to finishing on a nonlinear system is the ability to put in last minute changes. Here's an example of what this flexibility means in the real world. A major PBS series recently received a large donation from a new sponsor at the last minute. Using Symphony, the editors were able to shorten the program to accommodate the funding credits and show the producers multiple choices. On each of the four one-hour programs, they could see where those cuts would be and what they would look like even after the online was completed.

Another movement to get to faster-better-cheaper in the online process is toward streamlining the batch digitizing process. Avid has added in the

Symphony Timeline the choice to show offline media and the clips that are missing from the sequence (which are highlighted in bright red). As the digitizing process completes more of the sequence, each of the offline elements indicated by the red highlights begins to diminish, eventually leaving the Timeline color looking normal. This is a great way to keep an eye on your progress or to pinpoint a problem area if you think you have finished, but the display shows you otherwise. Offline media is also a good feature to use just before a Digital Cut since it will even show if nested media is offline. The user can follow the red elements down through the nested layers to the individual element that is offline.

TOTAL CONFORM

Total Conform is really the best reason to add Symphony to your production workflow. Although Symphony runs on an NT system, it takes all of the bins and projects from the Avid offline—which could be an NT *or* Macintosh system—and reproduces them exactly. If you have been using EDLs to conform your program to take advantage of the uncompressed quality of D1 linear tape suite or another NLE, then you have been working too hard. All of the effects that could not be represented by an EDL need to be re-created by eye. Every layer needs to be meticulously tweaked with effects, compositing, and drop shadows all started from scratch. Audio rubberbanding, pan and scan, motion effects, and AVX plug-ins are all automatically just there.

The rendering takes some time, but compared to having to make last-minute changes to a 15-layer composite on a digital disk recorder, which would you rather do? Every nested layer, every plug-in, and every keyframe of every effect is automatically carried over to the Symphony finishing stage through Total Conform. If the project was signed off on all the effects, then why does the client need to sit in the online to watch them tediously come together, one DVE move at a time in the linear tape suite?

With a little planning you can create templates of commonly used segments if you are assembling a program the same way every week. Just drop the new video inside the existing effect or composite and set it up to render. Now you can start multi-tasking, taking care of all the other responsibilities you have to do, like the pre-production meeting for the next project.

Symphony makes better use of expensive tape decks just by virtue of its being a nonlinear system. Your linear suite needs at least three decks most of the time, but with a majority of effects-intensive online compositing being spent without any decks rolling at all, this seems like quite a waste. Hang on, though. Don't try to take any of the decks away! Who knows when you might need them again? A nonlinear system uses one deck for a specific period of time and then doesn't need it again until it is time for the Digital Cut. You can take the deck to other Avid suites, for dubbing, viewing, and logging other projects.

ADVANCED COLOR CORRECTION

The newest version of Symphony includes a breakthrough in nonlinear color correction. The new Advanced Color Correction Mode sits alongside the other, more established modes of Source/Record, Trim, and Effects. Symphony's advanced color correction engine has a new interface and many new capabilities to address two types of users. The editor now has easy-to-use, powerful controls, and the more experienced colorists have enough of the established types of controls to feel comfortable. If you have done any color correction on another system for film to tape, desktop publishing, or on a high-end compositing workstation, you should see something that seems intuitive to use. Because of this wide, existing user base appeal, there will be very few people who use all the controls all the time, and many of the controls are meant to be redundant. If you find the one or two windows that perform the job you need, then look no further until you have a very difficult problem.

The Advanced Color Corrector handles two streams of uncompressed video in real time and still has enough bandwidth for another real-time effect and a DSK (either a title or a graphic with a static alpha channel). This mode is a primary color corrector and is designed to be highly efficient when matching shots scene to scene or making badly lit images look good. For more effects-style correction, you may want to use the Spot Color Effect that combines the vector-based drawing capabilities of the Paint Effect with the Color Effects controls. There are also eyedropper functions with NaturalMatch™ capabilities in this interface. Anything that requires field-accurate shape animation (keyframed moving objects) must be rendered, however, so even with the new color correction hardware, the Spot Color Effect is a software or blue dot effect.

If you are an offline editor, you may not be immediately interested in Advanced Color Correction, but perhaps you should give it another look. Many times there will be a discussion about whether a particular shot can be used at all because of on-location problems. In the past, you may have discarded a marginal shot or used it and crossed your fingers that someone else could make it work. So if you are lucky enough to be offlining on a Symphony system (or have quick access on a Unity Fibre Channel network), you may want to crack the manual. Give some of the graphical tools, like the eyedroppers, another try.

Technical Overview

The capabilities of the Advanced Color Correction Mode are a direct result of a new ten-bit RGB color correction chip that has been added to the Meridien board with Symphony version 2. The ten bits allow a wide range of adjustment with higher accuracy (1024 units vs. 256) and a very high-quality color space conversion from native YUV to RGB and back again. With many such conversions, there is a chance that the color in the new type of color space (YUV vs. RGB) will look different. The Avid system purposely minimizes this difference. There is no rounding of the color values, and there is, at most, a difference of one 8-bit unit (1 level out of 256) between the color in YUV and the same color in RGB. To further reduce the impact of this conversion, the system will pass the YUV signal all the way through the hardware without change if no adjustments require the extra power of RGB color space.

NONLINEAR COLOR CONTROL

What is it about nonlinear color correction that could possibly make it more efficient than a stand-alone color correction system? It's certainly not in the manipulation of controls. Working with the US$250,000 dedicated keyboard and control surface is more productive than using a mouse and a generic keyboard for color correction, but consider flexibility and efficiency of overall process when comparing the two approaches. First, when a dedicated color correction system is used in the film-to-tape telecine process, all of the original material is corrected and transferred. This may be hundreds of hours on long projects and requires the specialized skills of a full-time colorist and the most efficient interface money can buy. But don't try to do anything else in that suite!

On the Symphony finishing system, we have a different approach that views color correction as the last step in the process of a final master sequence. The editor is generally not responsible for correcting dozens of hours of footage and may even be doing a second pass on footage a colorist has already handled once. The pressure to crank through huge amounts of footage at an expensive hourly rate is not appropriate at this stage. Certainly, the Symphony system is not controlling a telecine device, so that stage will not be eliminated until everything originates from high-definition video sources (HDTV).

The Symphony color correction stage will more likely be used to replace tape-to-tape color correction. It will also be used to add that little extra value to projects that have originated on multiple sources like VHS, hi-8, stock footage, and different types of film. The overall evening of the images from scene to scene subtly helps to reinforce the message of the program by reducing distracting influences. Conversely, it can help to accentuate the differences if the transitions to different original sources is part of the storytelling. The color correction on Symphony can also be used to improve an inexpensive or quick "one-light" film transfer by adding contrast and saturation with a set of premade templates.

Using different types of source correction, the Symphony colorists can adjust large amounts of the program material at once. By choosing to adjust source color based on master clip or source tape, they can quickly affect every time that master clip or source tape is used. Although there may be some more fine adjustments required, this one step will help make up for some of the limitations of using a more generalized editing system to do color correction.

There is plenty of opportunity for skilled, professional colorists to participate and add value throughout the finishing process, even when color correction is handled nonlinearly. Their importance comes not from their ability to run a dedicated piece of equipment, but from their ability to instantly analyze, problem solve, and add input to the creative process. All of these areas of expertise can be translated to a nonlinear color corrector.

THE INDIVIDUAL CONTROLS

Although there is an entire manual devoted to color correction in Symphony, I just want to quickly outline the interface, what all of the windows can do, and offer a strategy of when to use them. The manual is clear and well-illustrated, so I highly recommend reading it after familiarizing yourself with this overview. As you look at the interface of the Color Correction Mode, you will see three screens across the top for scene-to-scene comparison. The center screen is the active image. With a right mouse click, any center screen image can be turned into a reference image, or you can use the reference image as the second half of a split screen. The split screen will default to the usual before/after comparison without the right mouse click.

Figure 12.1 The Center Screen Interface with the Center Split to Show Reference

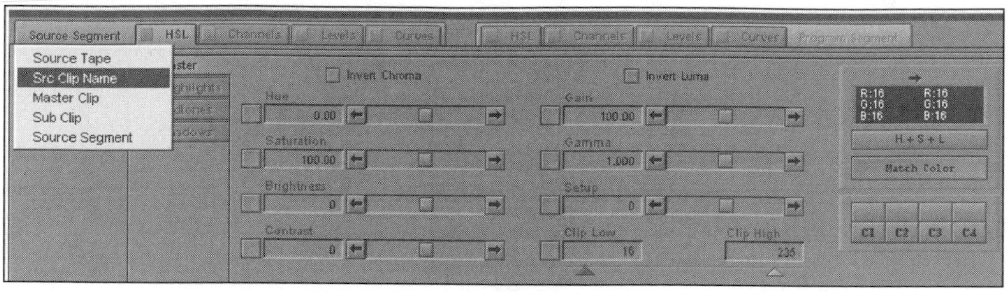

Figure 12.2 Different Kinds of Nonlinear Source Correction

Below each of the three screens is an individual Timeline for each of the panes, and the controls interface. Notice that there is a green half (the left) and a blue half (the right). The green controls apply to the source correction and the blue to the program, or Timeline, correction. Whenever you choose one side or the other, the colored lines in the Timeline will reflect this choice once the Correction display is turned on in the Timeline.

On the source side, you can pull down the menu to decide what source type will be affected. The default is segment, which restricts any correction to the active clip in the center screen. The other choices are source tape, master clip, and subclip. On the program side you can choose segment, track, and track "in to out" (this is a User Setting). I will go into more depth about these choices when we discuss the power of the nonlinear aspects of this interface. The main importance of having two separate sets of controls on the Timeline is that you can adjust each shot individually, in context of the sequence, as you compare scene to scene. Then you can add an overall correction to the entire sequence. The parameters of the two adjustments are summed, and one set of instructions is sent to the dedicated color correction chips so they do not create some kind of additive distortion of the image. This means that you can first make an adjustment that will make all the shots look even as the sequence plays, and you can then go back and add a creative "look" to a section.

Figure 12.3 Controls for Advanced User Configurations

There are lots of adjustments you can make to this interface to customize it. You can activate the controls you use most or optimize it for speed. The four icons in the upper right of the Color Correction Controls window give you direct access to a range of controls. The Correction Settings, the ability to save color correction templates, SafeColor Settings, and the ability to leave comments attached to a segment are all represented by icons. Comments show up in the Timeline and will carry through to an EDL. You can also save four different temporary templates by Alt-clicking the row of color stripe icons on the right of the interface. Normal clicking on the temporary templates will apply them, and Alt-dragging the Template icon to a bin will save it as a premade correction to be applied later and carried around with the project and bins. This is especially useful if a product shot needs to maintain a consistent look across a series of spots, for example, the red of a Coke™ can.

THE FOUR MAIN WINDOWS

The four windows of the interface are Controls, Channels, Levels, and Curves. The Controls window is what most editors will use immediately since it is close to what most basic color correctors do. Using the Controls window is the next step up in sophistication from the existing Color Effect. Hue Offset is the most powerful control in this window and perhaps the entire interface. Beginners can go straight to the Master Hue Offset, whereas experienced colorists will know exactly what to do with the other three hue controls for Highlights, Midtones, and Shadows. The luminance range will choose the part of the image adjusted with Highlights, Midtones, and Shadows. This can also be used as a very limited ability to do secondary color correction based on luminance values (somewhat like creating a notch filter with the EQ tool). The Controls window will be the most used and easiest to learn part of the interface for common correction.

The Channels window will be used mostly by experts or those most familiar with desktop publishing techniques. Rather than try to use this interface for color mixing or balancing, this mode is more appropriate for stealing detail and cleaning up noise. Desktop graphics gurus know that when a scan does not reproduce well, many times it is because one of the color channels lacks detail. With a little experimentation, you can find a channel with lots of luminance information, and that channel can be blended with a problematic channel. This

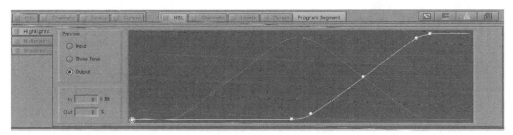

Figure 12.4 Luma Ranges

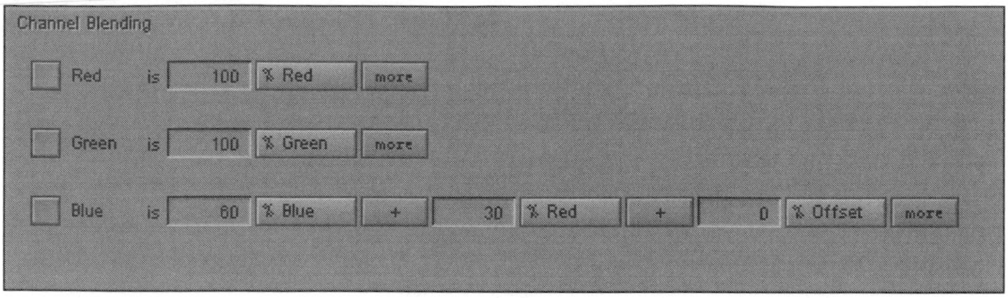

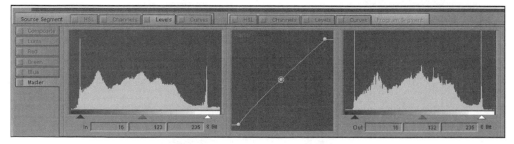

Channel Blending

Red   is   100   % Red   more

Green is   100   % Green more

Blue   is   60   % Blue   +   30   % Red   +   0   % Offset   more

Figure 12.5 Channels Window

same process can be done in the Channels window. First, view the channels to decide where the most detail may exist in an image. Then use that channel to blend with the channel that is slightly underexposed or overexposed. Using percentages of each channel, you can experiment until the overall detail of the image is restored.

The Levels window is the most advanced way to adjust luminance levels and the levels of each individual channel for optimum use of contrast. The histograms allow the user to see the distribution of energy across the range from dark to light. Two graphs represent the range of levels that will be affected: the levels that will be input into the processing, and then the final or output levels after processing. By graphically representing the energy of the image, you can see whether the blacks are all bunched together. They may need to be better distributed across the available range for a more even contrast. You can also see whether an overly bright image can be improved by moving the midrange or gray point.

The Curves window is the first window devoted completely to RGB adjustment. It shows the curves for each channel that can be manipulated with control points. Since this is a primary color corrector, you will eventually add more control points than required to improve the image. Try to keep the adjustments simple and you will have better results. This is also the main window for NaturalMatch, since that function works completely in RGB colorspace.

Figure 12.6 Levels Window

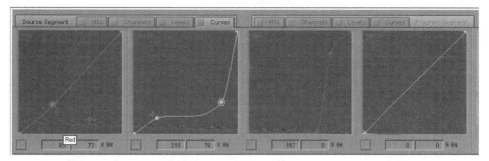

Figure 12.7   RGB Curves Window

NATURALMATCH

Eyedroppers are an easy and graphically intuitive method for matching colors from one image to another. The Eyedropper icon is available in many of the different windows in the Color Correction Mode. Usually, eyedropper controls apply specifically to the settings in that window. In other words, in the Levels Mode the eyedropper will allow you to pick the black point, gray point, and white point of the bad image and replace it with the corresponding points in the good image. This will extend the contrast range of a dull, washed-out image. The RGB values sampled with the eyedropper can be saved by Alt-dragging them to the bin, where they will be automatically named based on a database of standard color names.

Many times an exact match or absolute match does not return the results you really wanted. This is because you wanted only a part of the color values of the good pixels and didn't want the entire image to change based on those levels. What you really wanted was a relative match, or a match that gives results that are more natural. To address this, Avid invented NaturalMatch. This scheme of matching RGB values takes advantage of complex mathematics, which takes hue and chroma levels from the good image and applies them to the bad image. This makes an excellent starting place for more advanced matching, but many times it may be enough for the critical parts of an image to match from scene to scene.

NaturalMatch is especially useful for quickly changing a video-originated image that was not "white balanced" correctly. Pick a skin tone from a similar person in another shot and apply it to the skin of a non-white-balanced person (easily recognized as too blue or orange). The luminance values of the scene are maintained, but correct reference for hue and saturation is applied. The result is the right color in the target lighting condition.

SAFECOLOR

Another benefit of color correction in Symphony, from an offline editor's point of view, is the SafeColor™ feature. SafeColor guarantees that any level that goes through the video board will be corrected and brought within the specification created by the user and saved as a template. You can create safe levels templates for composite, luminance, and RGB levels and easily save them as Site Settings, or you can move them from system to system to ensure a level of consistency between editors and different edit systems, or as targets for different network requirements.

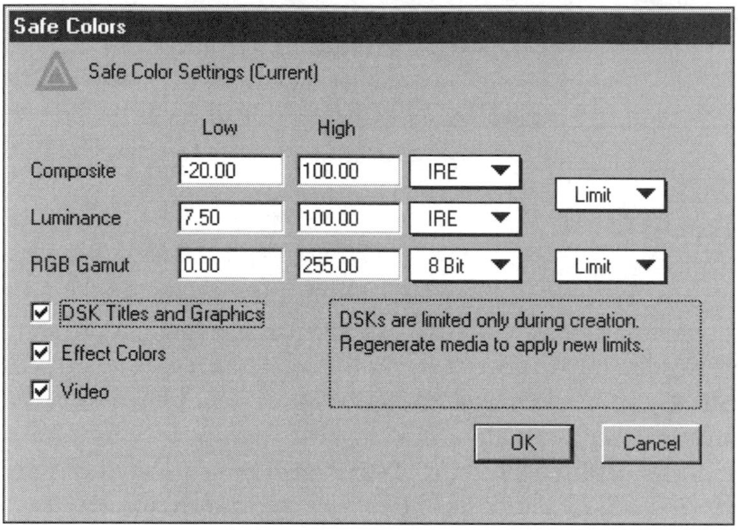

Figure 12.8 SafeColor Template

Any video that has been digitized or any graphic that has been imported without an alpha channel will be corrected using a clever method of pixel replacement. The standard system in most facilities to ensure broadcast standards is a processing amplifier, or "proc amp." Facilities may put a proc amp in at the output of their video switcher to make sure that all levels that go down on tape with illegal levels above and below a preset standard are chopped off. This "chopping" is sometimes referred to as a hard clip. It does not make the image look any better. In fact, a hard clip will cause all parts of the image above the preset level to lose detail. So where you may have seen a bright yellow object with shadow and edges, you will see a slightly darker yellow blob. Many of these proc amps will work to eliminate luminance or chrominance excursions above the broadcast standard. Sometimes the image will become darker, or less saturated. And sometimes you will lose all detail in the problem areas.

There is also a setting on some proc amps referred to as a soft clip, which starts to reduce the illegal levels at a preset point and gradually, over a wider range of signal, begins reducing the problem. This soft clip is less noticeable to the eye, but either begins to affect signals well before they are illegal or does not contain them completely within the legal standards.

The Avid Symphony system is smarter than this. Instead of compromising the signal, under many conditions, the Symphony SafeColor feature will improve the image quality while making a correction. It does this by identifying all of the pixels that are outside of the standard and replacing them with their closest neighbors that are within the standard. So bad pixels are replaced with the closest good pixels—many times adding back detail that would have otherwise been lost.

Does this mean that all editors unfamiliar with how to set a proper level on a shot will never have to learn? Of course not! That would be too easy and then anyone could do it! Setting the levels correctly in the first place will minimize the chance of illegal levels and will certainly produce better-looking images in the long run. Also, keeping the image within the digital realm of 16 and 235 (for both PAL and NTSC) will guarantee that your image is in the center of the optimum range for broadcast and reproduction. The security of SafeColors is to help those editors who may miss an occasional bright glint off chrome, the bright blue police lights, or other unexpected luma or chroma levels that are transient. Generally, online editors do not like to adjust the entire image for such tiny, but critical level problems. They rely on other tricks to keep the image within the proper range, but always optimized for the majority of the screen area. With SafeColor turned on, these problems will be fixed automatically.

The one potential problem to keep in mind is that there are two kinds of images that the Avid editing and finishing systems do not treat as real video: titles and imported graphics with alpha channels. Usually, they are treated as downstream keys (DSK). Both titles and graphics with alpha channels can be treated as a non-DSK by unchecking the DSK status in the Effects Mode, which turns them into normal video. Rendering any of these types of graphics will also bring them into the correct range since the rendered precompute is treated like standard video. If you want to keep them as DSK (the main benefit is that they will key in real time more often and remain uncompressed in quality), then you need to be careful when creating them. If the SafeColor limiting is turned on for graphics and titles, then that is the limit defined when they are imported or created. The Title Tool has a setting for safe levels where it will cut off the composite signal based on the safe levels template active at the time the titles are created. Imported graphics with alpha channels will also be limited to the levels active at the time they are imported.

If your levels requirements change after the graphics and titles have been created, then you will need to change the SafeColor template and bring it into the system again. Fortunately, this is easy. With titles, mark in and out around the entire sequence, turn on all of the video tracks, and re-create title media (it only takes a few seconds). With imported graphics that contain alpha channels, batch import them using the original files. Either of these answers could be tedious if you had, say, 20 sequences that needed to be fixed, but otherwise are fast. If you keep your level conservative to start with, then any change will likely be to make the requirements more liberal, and all existing graphics and titles will fall within the new tolerances. After all, it is rare to have your broadcast requirements change during the finishing stage.

USING NONLINEAR TOOLS

When the Symphony operator is going to create a source color correction, there are multiple, powerful ways to correct a large amount of material at once. The default setting for source color correction is Segment, which will restrict all corrections to the segment that is active in the center monitor of the three-window interface. The colorist may know that everything on that particular tape is slightly overexposed, incorrectly white balanced, or part of a "one-light" transfer that could use a premade template. A single correction or application of a template can then be applied to everything on that tape, master clip, or subclip. The same can be done on the Program side to correct an area that affects every clip between a marked inpoint and a marked outpoint. Check the Correction User Settings to make sure "Use marks for segment correction" is enabled (default is off), since it is considered a bit of a specialty usage with its own set of implications.

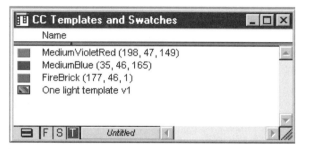

Figure 12.9 Saved Reusable Swatches and Templates

From that point on, there is a live link between the image on the screen and all of its subsequent uses. The power of a single tweak to ripple through a project is an extension of the power of a nonlinear system. If one of those changes sends a small part of one shot out of the defined correct levels, then SafeColor will kick in. It will quietly make the image conform without the user having to check out the entire sequence one more time.

Not all users are qualified to handle the power of live links, and one little tweak by the wrong person trying to make a shot look "cool" may result in widespread changes that are not desirable. You may want to take steps to ensure that the feature is used correctly. You can "flatten" the live links throughout the sequence so that all corrections default to being Segment only. This is done by loading the sequence and right-clicking in an empty part of the color correction interface. Choose Flatten from the right-click menu and feel confident about handing the sequence off to the next step in the process. Just in case, you may want to keep a version of the sequence hidden away in a secret place before you flatten it (or check Create New Sequence on the Flatten Choices window). There may come a time when the sequence is kicked back to you, and you will need to continue to make corrections with all of the live links intact.

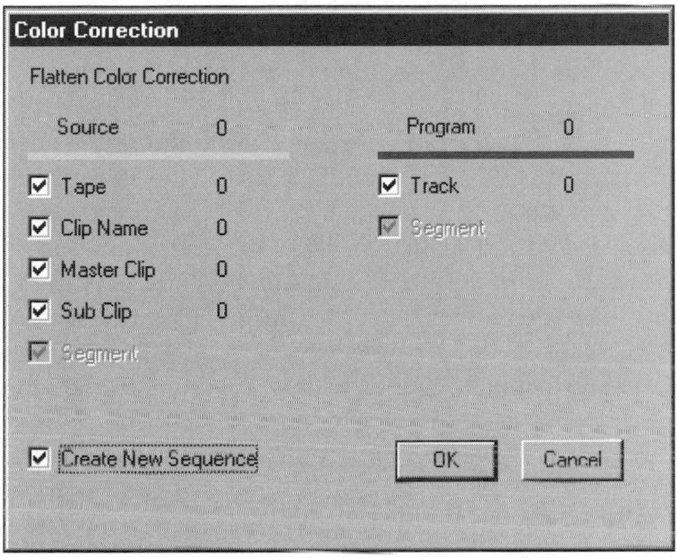

Figure 12.10   Flatten Dialog for Breaking the Live Links

Another tool in the nonlinear stage of color correction is Update, which takes any new shot added to a sequence and checks to see if it has been used before. If the shot has been used before and you want to apply the color correction added to it using one of the live links, you would choose Update. You can then decide if you want to update everything from the source tape, master clip, or subclip. If you have flattened the sequence, then all of the live links have been eliminated and Update will not make any difference. In the case of a flattened sequence, you will need to create templates and apply them to the new shots in the sequence.

The system cannot always assume that every time you want to add a specific master clip you want it to look the same. You may have added a creative look to the master clip the first time it was used (in a dream sequence or flashback, for instance). The next time the shot is used, it is meant to represent reality and blend in with the surrounding shots. This is why Update is a manual, controllable step taken by the user after a shot has been placed in the sequence.

SHARING MEDIA

With the growing adoption of media-sharing systems like Avid Unity MediaNet, there can be many people working on the exact same frame at the exact same time. This type of distributed, parallel workflow is revolutionizing post-production and is a definite step forward for nonlinear color correction. But even without a media-sharing solution like Avid Unity, there will be many people working on a single post-production job, in a serial form rather than parallel. The colorist or the online editor who is skilled in color correction can step into this process in many stages, and the Symphony Advanced Color Correction makes this as seamless as possible.

There is another control called Merge that, although it seems somewhat similar to Update, is meant to be used in a completely different set of circumstances. Imagine that while you are batch digitizing a project onto Symphony at an uncompressed resolution, your colorist is looking at the same images—the exact same media file—on another Symphony. While you are examining the shots and fixing dirt and scratches, your colorist is looking at the sequence for scene-to-scene correction. Perhaps while you are examining the sequence, your producer realizes she can make some changes, or some last-minute replacement footage arrives and you begin to make some new cuts. The colorist is busy color correcting on the original sequence, and the new sequence cannot be used until you are done with it.

What happens when you finish editing and the colorist finishes correcting and you discover the sequences are very different? This is when you Merge the color correction information from one sequence to the other. Take the colorist's version and the editor's version and merge only the color correction while leaving all the edits untouched. If there are conflicts where you have done some colorizing and you don't want it to be wiped out by the incoming changes, you have various choices for such conflict resolution.

Symphony version 2.1 is a first step toward the merging of nonlinear editing, media management, and professional color correction tools. We can look forward to uses that are more advanced and significant improvements in workflow as more editors and digitizers become skilled at color correction and take advantage of media sharing. Symphony addresses these situations through the live links, the ability to share correction information across multiple sequences, the security of SafeColor, and the straightforward controls for everyday needs.

WHAT IS 24P?

With all the discussion of how to best prepare for the coming of HDTV, many people fail to look at the most important aspect of all: the workflow. If you've read this far in *The Avid Handbook*, then you know that one of our most important design considerations at Avid is, "Does the new method improve workflow?" In a technological world where sales are made by trumpeting the next breakthrough that "changes everything," we often find that this method of advancement is one step forward and two steps back. If you look at the industry as a complex ecosystem, then some of these "breakthroughs" are like landfill for the wetlands. Sure, you can build the shopping mall, but what is the real cost?

24P is the emerging format for post-production that combines the frame rate of film with the quality and convenience of progressive frames. There is flexibility to use 24P for a wide range of aspect ratios and frame sizes for television broadcast, and an accurate film cut list for theatrical distribution.

The beauty of 24P is that it is fast becoming the most important aspect of workflow for HDT; however, it is not new. 24P is really an extension of the film work that has been going on for over 100 years and has a fully developed support system. Also of benefit to the Avid editor is that Avid was one of the first, the most successful, and the most recognized leaders in nonlinear editing in 24 frames per second. Although there is some new technology to get the progressive frames to work, the 24 fps part is an extension of Avid's existing Oscar-winning Film Composer system.

Taking into account all the possible different types of HDTV, there are 36 formats to choose from based on the tables published by the Grand Alliance (including all the different speeds). So which one will people really use? Already many networks have chosen to use a specific format that is somewhere in the middle of this range of choices.

If you had to pick one HDTV format that would be the easiest to convert into any or all of the others, you must take four important criteria into account. Consider image scan type, frame rate, frame size, and aspect ratio. What one format would convert each of these four items with the best possible quality into all other formats? If even one of these criteria is not adequate, then the final product will be rejected and the conversion will be useless.

Let's take a quick look at the issues surrounding each of these criteria. Remember that we are discussing origination formats, which may be very different from the final distribution format.

Scan Type

The requirement for standard-definition digital signals today is the ITU-R.bt.601 specification, commonly referred to as 601. This standard for digital images specifies an interlaced signal, and interlaced has been the standard for analog images as well. As we discussed in Chapter 6, the interlacing is a type of bandwidth and flicker reducer that was extremely important in the early days of television broadcast. Today, it causes many types of image degradation with computer systems that work in a progressive format. The interlaced image projects only a single field out of two at any one time. This is *half* the resolution of the image. Because of the persistence of vision and the latency of the RGB phosphors of the television screen, the two fields appear as a single frame.

There have always been problems with representing the film frame with interlaced images, especially with NTSC, so that you can have video frames with half of two different film frames. But much of our technology for editing, effects, and transmission assume interlaced frames, so any switch to progressive frames is a major undertaking. Having both fields onscreen at the same time with no temporal change between them will increase image resolution, which is what progressive scan promises. If you want to convert from one format to another, you will get a higher quality image when you go from progressive to interlaced than if you go the other way around. You are taking a high-resolution progressive frame (all scan lines present in a single field) and splitting it into a lower resolution, interlaced field. Because of the higher resolution, ease of frame-based effects, and interoperability with computer systems, given a choice of scan to be converted to all other formats, progressive is the better choice. This is the "P" part of 24P.

Frame Rate

One would think that the higher the frame rate, the higher the quality of the image would be, but this is only partially true. If you consider that the highest quality image most of us ever see is projected at 24 frames per second in the movie theater, then it is clear that there is more to quality than frame rate. There is some advantage to having a faster frame rate for sports coverage or other types of programming where motion effects are critical. If most people haven't noticed it at the movies (or from their video rentals), then this difference may not be obvious to them. Some broadcasters have decided to go with 60 progressive frames per second for very specialized needs. At this point, we are considering what would be the best universal choice for conversion during the transition to HDTV while still satisfying PAL and NTSC 601 broadcasts.

The main issues involved with choosing a frame rate have to do with converting between PAL and NTSC. Also important in the frame rate consideration is how accurate a cut list would be to cut film negative that has been edited on video and will be eventually projected in a theater. If you are making programs that have a wide distribution, the conversion from 30 fps (60 fields) to 25 fps (50 fields) has been expensive, time consuming, and never quite as good as starting the whole project over again in PAL. But, again, most people live with the compromises as long as the resolution of the images remains in the standard-definition state. If you want the best possible conversion, however, you need to take into account that the original PAL frame is larger than the NTSC counterpart (576 vs. 486 vertical scan lines). Converting from a larger frame to a smaller frame will look better than the other way around when working with pixel-based media. This means that the absolute best starting point for multiple-format output would be to shoot or transfer film to PAL. The continuing trend, however, will be to transfer film to 1080p (HDTV) and downconvert to 601.

As illustrated by the NTSC-to-PAL conversion issues, going from a high frame rate to a low one means blending or dropping extra frames. The best answer would be a lower frame rate that can be stretched with redundant fields from one format to another. Since the conversion from 24 fps to 30 fps or 25 fps is already commonly accepted, why not choose 24 as the universal origination standard? This also has the important side effect of being perfect for feature films that need theatrical release as one of the deliverable requirements. Starting at 24 fps means the possibility of a perfect film cut list using Symphony Universal or Avid FilmScribe™. Finally, because the conversion from 24 fps to everything else is done to the final sequence, you can be guaranteed that the pull down is consistent from start to finish. This becomes extremely important for high-quality, problem-free MPEG encoding. Since most network and cable broadcasts go through some stage of compression during transmission, this is an important time and money saver for the rest of the production process. Now we have the "24" of 24P.

Aspect Ratio

If you could just shoot everything on film or 24 fps HD, then all your frame rate problems would be gone and distribution would be easy, right? Don't you wish it were that simple? Not so fast! There are still two more important criteria to consider. The next is aspect ratio.

Standard television aspect ratio is 4:3, sometimes referred to as 1.33. The Grand Alliance decided on the most common alternative format of 16:9. This format was already being used in Europe and Asia for standard-definition digital betacam over the last several years.

If a program is shot in 16:9, many times the center action is "protected" for 4:3 or (just to confuse things more) 14:9. Protecting for these different formats means that the director of photography must make sure that all of the important action happens in the center of the frame. If the 16:9 master is broadcast as 4:3, then you don't want to piece together the story from sound effects as all the major action happens just outside the frame of your television set. Nor do you want to watch a conversation between two noses. On the other hand, the director of photography, having lots of horizontal area for compositions, needs to fill it with something!

Television programming is still primarily 4:3 and only very slowly will become 16:9 as more people buy home television sets that can switch between the two formats. A 16:9 television playing a 4:3 broadcast will have lots of empty black space on both sides. This is probably preferable to the letterbox look that relegates a beautifully framed picture to a tiny percentage of a small screen. Some DVD movies offer a 16:9 version for those who have the right monitors, but in North America today there is precious little 16:9 programming unless it is high definition.

When you shoot standard-definition television (SDTV) with 16:9, you are still restricted to the standard 4:3 frame size. The camera squeezes the image as it is being shot or telecined to tape. This is called 16:9 anamorphic (squished and funny looking) to distinguish it from 16:9 letterbox (black bars on the top and bottom). It is then unsqueezed during editing and playback so that those with the dual-aspect ratio monitors can see the original image. As you go out to tape for the final master, you output it in the original squeezed format for copying and broadcasting as 16:9 anamorphic.

The Avid editing systems can switch over to 16:9 by checking that mode in the Composer Setting or by right-clicking on the Source/Record monitors with an NT system. All titles, effects, and imported graphics need to be in the 16:9 ratio as well. The Avid system will create the proper sized titles when you create them in the Title Tool while in the 16:9 mode. If you forget to do this or change your mind later, you must re-create the title media while in the proper mode. As we shall see, the new Universal Editing and Mastering features in Symphony Universal have a slightly different approach and allow you to create both formats at the same time.

Effects are already created perfectly in 16:9 when you are in that mode with Avid editing systems. This becomes critical when you have circle or diagonal wipes. Keyframing of DVE becomes slightly different as well because you are composing the graphic positioning based on a different use of horizontal space. Be sure any compositing program you use for animations and advanced special effects can handle the correct 16:9 aspect ratio. You don't want to stretch or crop a 4:3 composition to fit!

All graphics must be created in the right aspect ratio when working in Photoshop or other graphics creation programs. Refer back to Chapter 6 for the proper methods for creating 16:9 graphics and importing them.

So if 16:9 is the new emerging standard and it is already possible to edit and finish 16:9 programs on the Avid editing systems, then how do you output 4:3? Again, consider that two of the most common answers, center crop and letterbox, are quickly becoming unacceptable to most viewers. The center crop would be to unsqueeze the 16:9 image and chop off either side in order to output as a standard 4:3. This is clearly not optimal for taking advantage of the original composition. This is why in Europe they protect for 14:9 because they figure they can actually broadcast to a public of mostly 4:3 television sets with a little bit of black letterboxing. They can also broadcast 16:9 without making any changes. This is a workable compromise if (and this is a big if) all of your material was protected for 14:9 during the shooting. This would not apply to theatrical feature releases that are broadcast. It is still not great for graphics and titles. Letterboxing has not caught on except for cinephiles who feel that the original image is too compromised when it takes up their entire 4:3 television screen. If the original composition is that critical, then 16:9 will not be good enough either. Most people still don't buy that a smaller image with big black bars is progress. But widescreen films of 2.35 aspect ratio will have less letterboxing when viewed on a 16:9 monitor, thus using more pixels for better picture resolution than viewing 2.35 letterboxed within a 4:3 frame.

The answer is a process called pan and scan. Up until recently, pan and scan was a very expensive session where the director sat in a telecine facility and chose the framing. Shots were moved and repositioned on a scene-by-scene basis. It could also be done in a tape-based online suite using a DVE, but this would still cost easily over US$600 per hour to keep the quality acceptable. No one likes to make creative framing decisions when the clock is running *that* fast. The expense and time involved meant that this technique was out of the question for many productions. That's a lot of money for future protection that, at some point, you will reuse the 16:9 version.

With the release of the Symphony 2 Universal model with Universal Editing and Mastering, and Media Composer 9.0 with Universal Offline Editing, pan and scan became a very simple, dedicated process using a real-time DVE. By using the Universal Offline Editing option on Media Composer, the director can sit in an offline Avid suite and choose all the framing and reposition moves at an offline rate. Once all those decisions are made, they will translate easily and completely to a Symphony system through Total Conform for uncompressed finishing. The pan and scan can still be adjusted at the Symphony stage if time is short and people will not make important decisions until the very last minute.

With the Avid implementation of pan and scan, the proper sized grid is displayed over the larger original frame and the editor can keyframe repositioning scene by scene. The editors can also create their own pans to include vital parts of the scene when the point of interest shifts with the actor's blocking.

You can also create a pan and tilt (or perhaps tilt and scan) where you make a 4:3 into a 16:9. This will help match legacy material with the new format. Since the 16:9 format will cut off the top and bottom of the 4:3 area, in order to change the point of interest, you create what appears to be a camera tilt.

As broadcast moves toward a 16:9 aspect ratio, you will need to intercut legacy material that only exists as 4:3 with new material that is 16:9. Now you can use the pan-and-scan tool to correct just the 4:3 material so that it may intercut with the 16:9.

Figure 12.11   Pan and Scan for 4 × 3 Delivery of a 16 × 9 Master

Frame Size

Frame size is another problem in dealing with HDTV. The problem is really rather simple: bandwidth. HDTV takes almost six times the bandwidth of SDTV. This means all existing systems designed for SDTV are inadequate and need to be significantly upgraded to handle all the items we have come to expect from an editing system. We are only discussing a single stream of HDTV and no audio! Imagine real-time effects and eight channels of digital audio with rubber-banded levels, DSK titles, and color correction!

Let's take a step back and look at the big picture. You will need to deliver HDTV at some point in the near future. For a chosen few, it is required today. For most, it will be in two or three years and, for others, perhaps even farther into the future. You can plan, archive, and shoot for it, but what are you actually required to deliver under today's tight deadlines? An SDTV 601 master. Perhaps even multiple versions of that 601 master, as discussed earlier.

If you work at 24 fps with a progressive frame, using pan and scan on a Symphony Universal system, you will be able to deliver all of your 601 requirements. Why would you make six SDTV versions in a US$1500-per-hour HD editing suite? Because you would need to do pan and scan and add titles. You are basically using the HDTV editing suite as an expensive downconverter.

With Symphony Universal you can output a 24 fps edit decision list (EDL) with the new standard as agreed upon with Sony. With such a list you will be able to go into any HDTV suite that uses the new 1080p/24sF (segmented Frame) HD decks and auto assemble the HD master. This is a bit of a reverse of the normal way of thinking, but why would you make the HD master, which is required sometime in the future, to be the primary goal? By starting with HD but downconverting to 601 for most of the work, every stage along the way will be cheaper, faster, and less prone to gaps in the HD support infrastructure. By the time you really need to deliver the HD master, there may be Avid solutions that will Total Conform in HD just as easily as going from offline to online on present-day systems.

CONCLUSION

The marketplace is changing very quickly in film and television. Formats are proliferating and standards are evolving. For those who need to stay at the forefront of the professional nonlinear editing market, it is important to understand what these changes mean to you and for companies like Avid to continue to innovate. Keeping an eye to the future, you can take advantage of Avid's leadership role and maintain the competitive edge in your market.

13

Simple Troubleshooting

This chapter is devoted to keeping you off the phone and on the system. There are many everyday situations where just a little knowledge of troubleshooting can keep you going forward, give you a bit more confidence, and, maybe, help keep your job. The first thing that can help you instantly is to read the Release Notes. This falls under the category of RTFM (Read The Manual, Please), but some people skip it because they just want to be up and running with the newest version no matter what. Stop, smell the roses, and read the known bugs. Avid is pretty good about listing what they consider to be the bugs you need to know. Your definition of a bug and their definition may differ, but you will definitely benefit by seeing that, for instance, one small part of something you need to do all the time doesn't work under certain conditions. It also helps to know if that procedure has been replaced by something faster, better, and simpler. Even if you haven't memorized the Release Notes, they can generate a little thought bubble over your head if something seems familiar. So don't toss the Release Notes; keep them handy and even scan them quickly before you call Avid Customer Support. Much of the advice in this chapter is based on the Macintosh since that is still the majority of the systems out there, but some of the information is not platform specific and may be useful to Windows NT users.

Customer Support has improved so much in the last few years that average wait times as of this writing are under four minutes and Avid takes it very seriously, but for you that's not really the point. There is time involved in figuring out you have a problem, realizing you don't know how to fix it, telling the other people in the room that maybe they had better get some coffee, and then dialing for help. Better to say something like, "Hmmm, did you know you are missing the active SCSI terminator?" and subtly imply, "Aren't you glad you hired me?" However friendly, competent, knowledgable, and good-looking Avid Support may be, you want to avoid talking to them until you have a serious problem.

ERROR MESSAGES

If you are not used to working on complex professional software, you may not be used to generating error messages. Write down the ones you see and, if they do not keep you from continuing, call Avid Support after the session and get the official explanation. Saying, "I got an error" isn't enough to let Avid Support help you figure out the problem. It may be something systemic or it may be operator error. But if it is a Fatal Error, you should be on the phone immediately if you don't know what caused it. After working on the system for a while, you will learn what causes the most common errors, and if you keep a log of when they occur and to whom, you can identify the pattern that applies to your own system, facility, and way of working.

The real trick to error messages is in deciphering them. They may be colorful but essentially meaningless until you figure out what the function is that has gone wrong or what kind of pattern they follow. A section of the programming code always generates a specific error message when something goes wrong. Essentially, error messages tell you what happened, not why. This is because the same error could have been caused for ten different reasons. The computer cannot look outside of itself and say, "You have too many unrendered effects at AVR 77 and the wrong terminator on Media drive 4A and that last sound effect on audio track 6 just put me over the edge!" It will say "Audio Underrun" because what happened was that it could not continue to play every frame of audio and video through that segment of the sequence. Many times this is the only explanation that the computer can confidently produce for why it cannot play.

THE IMPORTANCE OF CONNECTIONS

First, let's look at the basic world of connections. You may have enjoyed playing air guitar to Molly Hatchet, but air SCSI doesn't work as well unless it's connected to something and connected correctly. Connections are one of the first things to look at, especially if you have just moved the system. "I just moved my monitor and now it is broken," will lead most support reps to check your cables to the monitor.

Many of the connections to the computer have electricity running through them, and connecting anything "hot" can cause a component to burn out. A good rule of thumb is not to change connections while the computer is running (not including audio or video to an external source like a deck).

Cables that are screwed in tight don't come loose so quickly. This seems obvious, but many facilities decide it is easier to have them not screwed in so anyone without a Phillips head screwdriver can move things and make changes quickly. Do yourself a favor and buy a screwdriver with one end Phillips head and the other end flat. Tighten everything on the back of the computer and anywhere else you can tighten things down. Of course, if you yank really hard, you still have problems because now you have loosened the computer board it was attached to or even ripped the wiring out of the connector!

Make sure that things are fastened down away from big feet and spilling coffee, but don't pack them away so tightly that you can't get at them to take a look. Get a few bags of little plastic tie wraps from your favorite electronics store and wrap cables together in logical groups. You have to cut the tie wrap to pull them apart at some point, but better to discourage the unauthorized, and you can always wrap them again when you are done. The cost compared to down-time is negligible. Good engineers leave a little slack in all their carefully tie-wrapped suites so that any piece of equipment can slide forward enough to see around back. Large, easy-to-read labels that are also easy to understand make any phone call to Customer Support less embarrassing. "Is the video input connected?" "Uh, you mean cable VI649?" If you need to call your own video engineers, at least you will be sure that the problem is not software related. Also contemplate moving everything in the Avid suite a few feet away from the wall and mounting a small, clip-on lamp back there. This gives the system some more air if the room tends to get warm and will give you easy access.

MONITOR CONNECTIONS

If you have an Avid Xpress, you already know the benefit of working with high-resolution monitors. More open bins (if they each contain a small amount of well-organized material) keep you speeding along under certain projects. But more interesting is that when you are digitizing, you can have the waveform/vectorscope, the audio VU meters, several bins, and the console all open at once. On the other monitor of the Media Composer, the Record monitor, you can get a huge amount of real estate for your Timeline. Getting the Source/Record window with the two screens larger is a little tricky. You have to mess around with the cursor on the bottom right-hand corner until it changes shape and then drag the window so that it fills the screen horizontally. This positioning is saved with your User Settings so you only have to do it once.

There are two common problems with monitors. The first is getting the RGB and sync cables connected wrong in the back. How can you tell? All your colors will be horribly wrong. The monitor may be completely black if the sync cable is not connected correctly.

The second problem occurs only on the 20-inch monitors and has to do with the termination of the monitor. Your signal will be washed out and way too bright. Many people foolishly live with this because they didn't expect the signal to look great in the first place! They accommodate it by cranking down the brightness controls on the front of the monitor. To see this, look at the white type of the splash screens as you launch the software. If the letters are a little too bright, check the termination. Look in back of the monitor, near the input connections, for the termination switch. It has an icon that says 75 Ω, which stands for 75 ohms. Make sure your monitor is terminated since it is very easy to whack that little 75 Ω switch while moving or unpacking.

ADB CABLES

One of the silliest problems that seems to happen often is with the keyboard cables (USB is the newest version, but older systems use ADB or Apple Desktop Bus). First, let me caution everyone who is thinking of pulling this harmless-looking cable and reconnecting it while the system is running. You may have done this a hundred times, but the next time you do it, you can fry your keyboard or even your motherboard. Any cable that has power running through it has the capability to create a power surge or damage sensitive electronic parts by "hot swapping" or changing the connection while the power is running through it. The ADB connects directly to the motherboard, and the connection to the board is not replaceable. If that ADB connection fries, you need a new computer. If you don't know which cables have power and which do not, change all connections that are not standard video and audio only when the system is shut down. I know this is overkill, but simple and safe is a general policy.

The ADB cables definitely have power running through them. Unfortunately, ADB cables are always about an inch or two too short and so are under a certain amount of tension all the time. Many times just a slight pull is all it takes and your system appears to crash. You try all the keyboard reboot commands and nothing works and finally go to the CPU itself to hit the reset switch. When everything comes back on line, surprise! You still don't have control because what really happened was the ADB cable had come just slightly loose.

One last consideration with the ADB is how many devices to attach in the chain. Many people like to use trackballs because they don't have a lot of space to manipulate a mouse. They also like the programmable buttons for repetitive or common functions. For those who don't like the feel of a trackball, there are programmable mice now, too. And the most versatile device to connect to the ADB chain is the graphics tablet. Instead of suffering the effects of the mouse or trackball death grip at the end of the day, many people like the feel of a light, familiar pen. Even if you find that you don't like editing with the pen, it is the best way to do graphics. Simple graphics, like fixing a scratch or a dropout, becomes faster and easier when you get away from the "painting with a bar of soap" feel of the mouse.

SCSI CONNECTIONS

The ADB connection is basic and straightforward compared to the scariest of all computer connections — SCSI (Small Computer System Interface). The term SCSI

Voodoo may not be completely foreign to you and for a good reason. Even though you may follow all of the complicated rules of SCSI or simplify your system to minimize them, you may still encounter situations that just don't make sense.

Basic SCSI Rules

There are some basic rules to keep in mind no matter what your SCSI configuration is. First, all media drives that are playing video at any resolution above AVR 3 should be connected to the SCSI accelerator card and not to the other SCSI connection that comes built into the Macintosh. That other connection, away from the other cards, should be used for devices like slower drives for graphics, backing up, removable media, scanners, or the external 3-D effects "pizza box" (the Aladdin). Avid, like many professional software companies, is forced to complicate the SCSI situation because standard SCSI connections are not fast enough to pass the huge amounts of video and audio data at the speeds required for real-time playback at high resolution. Avid adds an extra card to the Macintosh called a SCSI accelerator. The SCSI accelerator is made by a company called ATTO, and this is where you generally connect all your media drives. The more things you have connected at once, the more complicated your SCSI troubleshooting will be.

Second, keep the length of your SCSI connections as short as possible. The cables that ship with your drives are meant to be that short because anything longer than the maximum length of normal SCSI chain causes serious voodoo behavior. There are now four supported types of SCSI drives: Classic SCSI-1, Fast and Wide SCSI-2, Ultra SCSI-3, and Ultra2 LVD (a variation of SCSI-3). Each type uses different cables and different cable lengths. So rather than try to outguess the manufacturer, stick to the length that comes with the system. The length of the cable used for the entire SCSI chain must take into account the length of cable inside each drive. That can add up pretty fast at over a foot per drive. A common mistake is to take one look at the length of the cables that came with the drives and rush out to the electronics store and get the longest cable you can find. Avoid the temptation to use these and keep the cables short.

Make sure all your SCSI cables are of the same brand, type, and style—different cables may have different internal configurations that may cause some devices to just never work right. This is why it is so important when mixing different types of drives to get all the cabling and termination correct, as we will cover later.

SCSI cables are very sensitive to twisting and must be handled more carefully than any other cable on the system. Just bending a SCSI cable back and forth a few times can significantly reduce its functionality. Don't strain, kick, stomp, or bite them.

Third, turning drives on and off has its own set of rules. Make sure all SCSI devices are turned on and have clearly come up to speed before powering on the Macintosh. Listen for each drive to make a single "click" sound when it has finished spinning up. Keep all SCSI devices on if they are connected in the chain to ensure consistent behavior. Turn all peripherals off only after the Macintosh has shut down.

There are lots of complicated rules about SCSI, especially when you are combining the newer wide drives with the older narrow drives. These three terms, fast, wide, and narrow, refer to the speed and capabilities of passing larger amounts of data through the SCSI chain. The drives themselves don't look all that different. Fast and wide drives (the iS9 and iS23 drives, for example) are the minimum kind of drive supported for two-way striping to playback AVR 77. You can use the older four-gigabyte and nine-gigabyte drives for AVR 77, but you must use a dual ATTO card or two ATTO cards and stripe four of the same size drives together (two on each ATTO card). This is called *four-way striping* and is like having one really fast drive with four read/write heads. Four-way striping is a good way to continue using older drives, but if one drive fails, you will lose all the data on the other three drives, too.

There is yet another kind of SCSI 2 on older Macintosh editing systems: SCSI-D or differential SCSI. Differential SCSI is no longer a shipping configuration, but you may find it on older systems, especially Film Composers with SCSI MediaShare. The differential SCSI is for drive towers and storage expanders that allow you to connect many more drives than either the standard seven or eight and are capable of the much longer cabling lengths of 75 feet (compared to 18 feet for single-ended SCSI). The important thing about a differential connection is that it has a non-compatible connector to keep you from accidentally connecting regular drives and burning them out because of the difference in the amount of power flowing through the cables. Don't confuse MediaDocks with drive towers. MediaDocks do not require the differential ATTO card or the differential cables.

Termination

All SCSI chains must be terminated. Termination is the way the computer knows where the end of the chain is and which is the last device. It keeps the signal from bouncing back to confuse the computer with false signals. Technically, the chain must be terminated at the beginning and at the end, but on most Macintoshes (all the newer ones and all the ATTO cards), the termination at the beginning is internal. This means that you need to attach the terminator that comes with the drives to the last drive in the chain. Don't worry about termination at the beginning of the SCSI chain (although occasionally that internal termination can fail, too). Always be sure to use an active terminator, the purple terminator that came with the narrow drives or the blue one that came with the termination kit for the wide drives. With the new LVD drives the terminator is beige and has an LED that indicates whether it is in the LVD mode (green) or single-ended (amber). We'll discuss the difference when we deal with mixing drive types. Don't use the generic gray terminators from the electronics store—they are generally not active.

SCSI ID

The most basic SCSI rule is that the number of the SCSI ID, the number associated with this drive in the chain, must be unique to the chain. The SCSI ID is set on the back of standard drives with a pen or other pointed object; in MediaDocks it is set on the front panel. Before you attach any drive, make sure that its ID does not repeat a number already being used by another drive on that SCSI chain. The fact that this fundamentally important piece of information is a small number on the back of the drive again points out the importance of giving yourself access to the equipment after it is installed. How do you know which numbers are being used unless you can stick your head back there with a flashlight and read them upside down and backwards? The best way may actually be another piece of software—the Avid Drive Utility (ADU) on the Mac or the Disk Administrator on the NT. These utilities show you the correct SCSI ID even if the ID number on the back of the drive is broken and displaying a wrong number! Of course, you need to be able to boot your system with the drives attached in order to use this software. The new LVD drives finally have the SCSI ID on the front of the drive.

Getting the wrong SCSI ID may cause the system to not boot correctly or even damage data. The Macintosh desktop may also indicate that you have many more drives connected than you really do. When there is a problem with the SCSI chain on startup with the Mac, you may get the flashing question mark icon. This is because the system cannot find the startup drive with the System folder on the SCSI chain. You have somehow confused the SCSI chain that has the internal Mac startup drive.

Take into consideration that some devices come already terminated internally and must always go at the end of the chain, like the "pizza box" external 3-D DVE. If you have two such devices that terminate internally, they are going to fight with each other until you eliminate one or figure out how to unterminate it. Always shut down the system first if you are having a problem with SCSI IDs. Do not try to change the SCSI ID while the system is running!

With narrow drives, you can use only the numbers 0 through 6 for a SCSI ID. A wide drive can use 0 through 6 and, if connected to a PCI host computer (Macintosh or NT), 8 through 15 can be used as well. Never, ever use SCSI ID 7 on any SCSI device since this is the number used by the ATTO card itself (which is also technically a SCSI device) or by the host computer. On the internal Macintosh SCSI chain, avoid SCSI ID 0 because that is the ID used by some internal, system drives. Internal CD-ROMs are generally set at the factory to use ID 3, so when attaching devices to the Macintosh SCSI chain you should avoid SCSI ID 3 as well. Scanners and Iomega Zip drives may use SCSI ID 5 or 6. When you are adding and subtracting drives from any system, the ID is the most important factor in making sure the drives work happily together.

Connecting Wide and Narrow Drives

We have already discussed the difference between narrow and wide drives and, when you upgrade to wide drives, you need to figure out how to connect them to the same SCSI chain with the narrow drives. The narrow drives must come last in the SCSI chain.

When connecting wide and narrow drives on the same chain, you need a special termination kit from Avid. Wide drives use a 68-pin connector and narrow drives use a 50-pin connector. The difference is more than cosmetic, however, since there are more active pins inside the 68-pin connector, and these extra pins must be terminated before connecting to the narrow drive. You get a new wide cable to go between the last wide drive in the chain and the first narrow drive, a small blue terminator/adapter that allows this cable to connect to the narrow drive, and a blue terminator to go onto the last drive in the chain. Use the standard wide cable (68 pin to 68 pin) to go from the Macintosh to the first wide drive. Then use the termination kit cable (also 68 pin to 68 pin) with a special blue terminator adapter (68 pin to 50 pin) added to it from the kit when you connect the wide and narrow drives together. Finally, use the special blue terminator on the last drive in the chain.

Cables to connect the wide ATTO card to the narrow drives (68 pin to 50 pin) are very different from cables that appear similar for connecting wide drives to the narrow drives (68 pin to 68 pin with a 50-pin adapter/terminator). Even though this narrow cable may connect, you will have nothing but problems connecting wide drives to narrow drives. It is a good idea to label these cables as for narrow use exclusively. Better to lock up the 68-to-50-pin narrow cables somewhere after upgrading to wide drives.

Connecting LVD Drives

LVD are the newest SCSI drives offered by Avid. They are twice as fast as Ultra SCSI and four times as fast as Fast and Wide SCSI. They can play back 1:1 resolution with a two-way stripe across two LVD disk controllers. LVD drives require the same 68-pin cables as the Fast and Wide, but if you connect an LVD drive to a SCSI chain with a slower drive, then they drop their speed to match. In other words, an LVD drive that is connected with an iS Pro drive will only perform as fast as the iS Pro.

There are still many complications to the SCSI chain, but these are the basics and must be taken into consideration on an everyday basis. Even if you follow all these rules, you may find that a particular drive works on ID 4 but not ID 2. You may never find a good reason for this (although there is a reason). You may also find that the order of the drives in the chain makes a difference even though all of the termination is correct at the end of the chain. And above all, you may have to juggle extra devices on the internal SCSI to find the best order. Slowly rebuilding the SCSI chain one device at a time and rebooting is often the only way to isolate where the problem is occurring and identify the problem device, cable, or terminator.

MOVING STRIPED DRIVES ON NT

For all of NT's high-speed, memory-efficient, stable architecture, it still lacks an easy way to move striped drive sets from system to system. We hope this will be improved in Windows 2000, but in the meantime we must take some very serious precautions to make sure the striped set will show up and work correctly on the second system.

Stripe and format the drives following the User Manual using the NT Disk Administrator. You will need to have Administrative privileges to do this. You cannot have any extended partitions to stripe drives, so you may have to delete those partitions before you proceed.

- Run the Avid-supplied software Disk Mounter.
- Click on the Register button.
- Click Do It to confirm the procedure.
- Exit Disk Mounter.

Information about the striped set is stored on the drives. This information will be important to the registry of the new system when they are mounted. Make sure that the drive letters or the SCSI ID of the striped drives will not conflict with the drive letters on the second system. If you don't know what drive letters are already being used, then change the drive letter of your striped set to something high in the alphabet.

Check to see if the original striped set used two SCSI controllers (like the LVD drives). If they use two channels, then they must be matched up to the same channel on the new system. If the first drive was connected to the A channel of the SCSI controller, then it should also be connected to the A channel on the new SCSI controller. You may want to label the outside of the drives to make sure they are matched correctly by the person making the final connections.

After the striped drives are connected to the second system, then you should run Disk Administrator to make sure the drives are recognized. Then run Disk Mounter and use the Mount button. Restart the computer after mounting the new striped set.

Occasionally, you may be trying to mount striped drives on a brand-new system that has never been used to stripe drives. You must activate a Devices Control Panel to recognize the striped file system for the first time.

- Go to Devices Control Panel under the Start menu and Settings.
- Find the device driver Ftdisk in the list of devices in this window. If this is your problem, then the Ftdisk is probably disabled.
- Highlight Ftdisk and click on the Startup button.
- On the Startup Type window, choose Boot and OK.
- Back at the Devices window, click on Start.

You may have to reboot the computer for the changes to take effect. If not, then you should now have access to the striped drives.

Another problem with drive registry data is that it may become corrupted on the second system and not allow you to see the new drives. You will have to delete the corrupted registry data and create new data. First run Disk Mounter and register the data for all the drives before you start this procedure.

- Go to the Start menu and choose Run.
- Type "regedt32" and OK to get the Registry Editor.
- Find HKEY_LOCAL_MACHINE and open this window.
- Open the System folder.
- Highlight the DISK folder and use the Delete key to delete it.
- Say yes to the Warning and close the Registry Editor.
- Open Disk Administrator again and allow it to write a signature to each disk.
- Use Disk Mounter again to mount the drives and say OK to each striped set.

Never try to hot swap drives on an NT whether they are striped or not. NT will not see new drives until the system has been restarted.

AUDIO CONNECTIONS

You also need to be concerned with audio connections. If you are doing anything with multiple audio sources, consider a small, inexpensive mixer. Check to see if your audio card is an Audio Media II or an Audio Media III card. As I pointed out in an earlier chapter, be sure to use the Avid-supplied audio cables going from any source, like a mixer, directly into the audio card. A built-in attenuator in the cables makes sure that the level going in matches the level going out. You cannot compensate for the lack of these attenuating cables unless you have a mixer that outputs at –10 dB.

If you have self-powered speakers, then connect the speakers to a power strip with everything else and turn them on through that strip. This is a good idea for most of the equipment since it lessens the possibility that a piece of equipment will be left off accidentally. If the speakers are on and you still have no sound, check the Audio Tool (Alt/Command-1) to make sure you actually have audio playing back. Change the Timeline view to show Media Offline and see if the segment lights up bright red. If you have an older system, you can change the Text in your Timeline to show the media file name and see if it says Offline. If it gives you media file numbers in the Timeline audio tracks and you still can't hear any sound, then check your Pro Tools external hardware. Make sure that the sample rate is set to the same rate the audio was digitized at. This can be confirmed with a heading in the bin or by looking in the Console, which will clearly tell you that you have a sampling mismatch. If the sound was originally digitized at 44.1 kHz and the switch is set to 48 kHz, you will hear nothing. Also make sure you are set to .99 or 1.00 to match the original digitize setting. This external switch can be hit by accident or when switching back and forth from a film project that may need the .99 pull-down rate. If you need only one type of sampling rate all the time, I suggest using a little piece of black gaffer's tape to permanently prevent these switches from being accidentally changed. More recent systems have this change of sample rate as an internal software switch in the Compression Tool.

THE BLACKBURST GENERATOR

A blackburst generator (BB Gen) is like a synchronizing clock for video or audio. It provides a steady source of a perfect video signal: black. I highly recommend one of these even if it doesn't come with the system you ordered. You can connect it to decks, monitors, your video card, and the Pro Tools video slave driver to synchronize your audio. Many decks require a stable signal to their composite video input when they are playing back, especially the Sony PVW-1800. If you are using the component inputs and outputs, consider connecting blackburst to the composite video input permanently. It makes it easier to black tapes when you are not using the deck for anything else.

You should connect things as permanently as you can and design a system that requires as little connecting and disconnecting as possible. Consider patch bays, mixers, MediaDocks, networks, and removable media like Zip or Jaz drives. The less wear and tear you put on cables, the more reliable and long lasting they will be. A little more initial investment during the planning stages can positively reduce troubleshooting downtime in the future.

STANDARD COMPUTER WOES

Even if you successfully eliminate the potential for problems with connections, there is still the potential for standard computer-type problems. Unfortunately, these kinds of problems are not easily solvable by the typical MIS department, even if you are lucky enough to have one. This is one of the reasons it is a good idea to be able to take care of your own computer problems. Standard computer support personnel are going to be at a loss with the Avid problems unless they have been through some training for the specific requirements of high-resolution video.

Extension Conflicts

An extension is a small piece of software that launches when you start the computer and runs in the background. They are necessary for the operation of certain software applications. The Avid systems have a handful of extensions that are installed automatically in the Extensions folder in the System folder when you use the installer disk. You can run Adobe Photoshop, After Effects, and some other third-party programs that require their own extensions on your edit system computer without problems and, while many times this is efficient and practical, it is sometimes asking for trouble.

If you use programs that add extensions when you use their installer, eventually you will have an extension conflict. Sometimes the conflict is obvious and you cannot boot the system and run the Avid software after you have installed something. That problem is pretty easy to fix. In fact, there is a program on the Mac called Conflict Catcher by Casaday and Green, Inc. that systematically cleans out your extensions folder and puts extensions back in a way that allows you to quickly zero in on the culprit. You may find your life simplified by making several sets of extensions and labeling them for the particular functions they are used with. Have a set of extensions for graphics work, one for cruising the Internet, and one that is stripped down for editing. The best set of extensions to create is one that allows you to work with the three or four most useful programs that you run all at once. Test to make sure there are no conflicts, strip the extension folder so that only those applications are supported, and then save it as an extension set. (Back it up in case you are forced to reinstall your operating system.) Your system will run faster on less RAM and give you many, many fewer mystery crashes.

The most difficult extension problems to track are intermittent or related specifically to particular functions. They may occur only when you go into the Digitize Mode or only when you are making a Digital Cut. These may occur because some third-party application has changed some important function in your system. Another video application may reset the frame size of the video board or change the media drive firmware. There are many small, but important settings that may need to be reset after running another program on the Macintosh. You may only be able to find these programs after they cause serious problems, but if those problems are intermittent, you may never connect the two as cause and effect. Simple is better, and if you can afford another Macintosh to run third-party applications, you can reduce the amount of crashes when you are running the Avid software.

There is a set of general rules that you should follow when contemplating adding any software to your editing system that is not directly related to editing on the Avid. Most of these rules revolve around the requirements that the Avid system has to be uninterrupted during critical operations like digitizing, playing, or rendering. Try to close the Explorer window on NT since it will attempt to continuously update in the background. In general, don't leave other applications open that might run background processes, like Microsoft Office. The Find Fast background process is another common application that will try to update in the background. Some e-mail applications will periodically check the network for new mail and may interrupt some other more critical function. Calendar programs are just as guilty of trying to schedule and update at preset times. Screen savers can start to play at inopportune times and may cause the system to hang. Keep your system fonts small and do not use large graphics for the desktop wallpaper display. Keep your editing system primarily for editing and, unfortunately, many of the things a user can do to customize the computer, the fun stuff, may also pull valuable resources away from functions that are already pushing the limits of the system.

Working from a Floppy

An organizational error that is unfortunately too common is trying to work directly from a floppy disk. Do not ever open a project, a bin, or a User Setting directly from a floppy. The moment you remove the floppy, but continue to work with the bin, you will have very strange errors. The Avid system periodically needs to save back to the original bin file on the floppy. Even worse, you will not be able to close the project or save the bin you are working on if it grows too big to store back on the floppy. All bins automatically save themselves when they close or when the project is closed. You must allow for the extra space if the bin has grown larger.

Access to Original Software

One of the final solutions to difficult, intermittent, or unusual problems is to reload the software. Avid Support may ask you to reload the Avid software, the Macintosh or NT operating system, or both. All users of the system should have access to these disks. A big mistake, commonly made, is to lock these disks up safely away from anyone who might need them, probably at 3 A.M. Make sure the disks are the most recent and correct versions of both the Macintosh operating system and the Avid software. You may cause more problems by loading an unapproved version of the Macintosh operating system.

Access to the Hardware

None of these techniques do you any good unless you actually have access to the computer itself. Some installations have cleverly hidden the system away in another room or rack mounted it in a machine room. If you are completely forbidden to touch the hardware because of facility or union rules, then forget about it. You can just hand the phone to the appropriate authority. Otherwise you must, phone in hand, be able to look around back and see that all the cables are tightly connected, that the power to everything is on, and what the disks' SCSI IDs are.

All of this means that you should look into getting a good engineer to set up your suite. Good engineers are worth much more than their salaries when they save you the embarrassment of your first several jobs going out the door with bad levels because you were monitoring the audio in the wrong place! Make sure, before the engineer leaves, that you have a thorough understanding of the cabling and get a wiring diagram you can refer to.

Never Enough RAM

More RAM is always better, but there is one thing you must do after buying and installing (or getting a certified Apple Technician to install) your RAM. On the Macintosh, you must change the allocation of the application. Find the icon of the original Media Composer software and click on it once. Then choose Get Info (Command-I) from the File pull-down menu at the Macintosh Finder level. Change the minimum and preferred amount of RAM to take all of the new RAM you installed except for about seven to ten megabytes for small programs and enough for the Macintosh operating system to expand when it needs to. To be really sure you are not shortchanging the operating system, go to About this Macintosh under the Apple Menu in the upper left-hand corner of your screen while you are at the Finder level. See how much RAM the system is using now and add a few megabytes to it. Subtract that amount from the entire amount of RAM available and give it to the Avid application. If you have a large amount of RAM, well over 100 megabytes, and you don't have a huge amount of media online, over 200,000 objects, then you might want to leave a larger chunk available for a graphics program or EDL Manager. With versions 7.x/2.x and later, you also have the DAE (Digidesign Audio Extension), which will launch the first time you use an Audio Suite plug-in. This extension may take up to 20 megabytes of RAM. EDL Manager for very large sequences may need 20 megabytes as well, but most of the time it should work fine with around 12 megabytes.

STARTING UP FROM ANOTHER DRIVE ON THE MACINTOSH

One of the first things you need to do if you have problems when you boot your system is to try to start up using another drive. It is easy enough to bypass all your extensions if you think that is causing the problem by holding down the Shift key during the initial moments of the startup. But if the problem is truly with the internal Macintosh drive, you can force the Macintosh to look for another drive with a System folder on it and see if you can boot at all. Once your system is up and running, you can try to track down the extension conflict or run a disk repair utility to try to fix the problem.

The Macintosh looks to see if there is a floppy disk with a Startup folder already in the floppy drive before it looks to its own internal drive. This is where Norton's Emergency Disk comes in handy. It contains a stripped-down System folder and Norton's Disk Doctor, which you can then launch from the floppy and use to repair the internal drive. If you don't have an Emergency Disk, you can force the Macintosh to look for another drive before the internal drive, like the Disk Tools floppy that came with your system software (if it came on floppies). You can boot from a CD-ROM that holds the installation of the Macintosh operating system, too. If you simply hold down "C" while booting, you will default to starting up from the CD-ROM. If you hold down Command-Option-Shift-Delete, you can boot from a CD-ROM, a Zip, or a media drive if they have a viable system folder. I carry a Zip drive and 100 megabytes of utility programs with me. I also carry a clean System folder that I copied from a properly functioning Avid system.

Again, having access to a clean Macintosh operating system and the CD-ROM of the original Avid application may be an easier and faster answer than really trying to pinpoint a problem. You can always go back and try to re-create the problem when you have more time and no one is threatening you with deadlines. Some facilities that regularly work under tight air dates will keep a mirror image of their internal drive on another external drive attached to the Macintosh SCSI chain. If there are any internal drive or system-related problems, they can force the system to boot from the external drive and be on their way. Another quick way to deal with internal system problems is to do a clean reinstallation of the Macintosh operating system. This is as easy as loading the system software installation disk and holding down the keys Command-Shift-K (for klean, I guess) while rebooting. This automatically puts the old operating system in a folder called Previous System Folder and installs the system software again. With version 7.6 and later, there is a checkbox for the clean install since Apple finally realized how important this operation is in a pinch.

ROUTINE MAINTENANCE

By far the most important day-to-day steps you can take to keep your system running and happy are just routine maintenance. These techniques are so fast and easy that as a freelancer, I used to come in a little early on the first day of a project to run them on any new machine before I started.

Rebuilding the Macintosh Desktop

The first procedure is called rebuilding the desktop. Do not underestimate the power of this procedure. You are telling the Macintosh to look at every file on all the drives and build a new database of exactly where everything is. There are so many important pieces of software that the Mac must find instantly to run smoothly that having everything accounted for before you begin can prevent all kinds of nasty little surprises. When you start up in the morning, or if you are restarting because you have had a problem, hold down the Option and Command keys during the startup process. It takes longer to start up as the machine looks at everything on all the drives, but it is worth the wait in the long run. You don't need to do this every day, but if you have been moving, deleting, or adding large amounts of files, then it's like chicken soup: it can't hurt. You can rebuild the desktop using Norton's Disk Doctor, which we will discuss shortly. When in doubt, ask Avid Support about versions of any third-party programs that you use on your system for maintenance.

Disk Diagnostic Programs

The next important step of routine maintenance, a disk diagnostic program, takes a little longer, but if run once or twice a week, maybe more often in a facility that never sleeps, it allows you to catch serious problems before they become fatal. Disk diagnostic and repair programs have been around for a long time, and people may have had a bad experience with them and media files, but they are a lifesaver to run on internal drives. Norton Disk Doctor is the most popular and ships with the Avid Macintosh systems. On the latest versions of the Avid Installation CD, it is in the Utilities folder with just the parts that you should use. In the past you may have received the entire shrink-wrapped Norton Disk Utilities package. If you did a full install of all the options, you installed a few extensions that can interfere with the operation of the Avid system. All of this correct installation information, of course, was in the Release Notes, but . . . just install the Disk Doctor and leave the rest for a Macintosh that doesn't need to play back gigantic video files with real-time effects!

There have also been the rumors that you should never run Norton Disk Doctor on media drives. Part of the problem was that in trying to repair a gigantic media file, old versions of Disk Doctor occasionally left the file corrupted. More recently, the controversy was over the fact that Disk Doctor couldn't deal with striped drives. With Disk Doctor version 3.2 and later, that is no longer true. I run Disk Doctor on the media drives now, but with one exception: I never repair a media file. If Disk Doctor wants to repair header information and other minor directory problems, it can go right ahead, but I don't want it to mess with my 100-megabyte media files.

Speed Disk is a utility that Norton ships with Norton Disk Utilities that defragments, or reorganizes, a drive so that there is no wasted space. It optimizes disk space by moving files around and, in a very orderly way, filling in gaps on the drive. This is great if you are concerned about space, but not so great if you are concerned about performance of gigantic audio and video files that must play back simultaneously. Avid has determined how to record files to the drive in such a way that video and audio stay together to maximize efficiency when the drive head is reading the file. That way, the drive doesn't have to work so hard at the high speeds that are necessary to keep it from underrunning. If you digitize audio and video to separate drives, this is not such an issue, but many people don't have that luxury. It is standard procedure for systems with fibre drives to digitize audio to another SCSI drive. Note that if you put audio and video that are meant to run simultaneously on different partitions of the same drive, you are forcing the drive to work even harder. With the popularity of striped drives (and the repeated assurance that you will not lose half your drive space), digitizing audio and video to different drives is less of an issue. Are you really pushing the system, trying to get many real-time effects at high resolution for instance? If you are pretty organized, then you may benefit from splitting your audio and video to separate drives or even separate SCSI cards.

NT systems come with a utility called CHKDSK, which will check a drive for problems. You can get to CHKDSK in the Disk Properties window. Right-click on the drive icon to get the drive properties, then choose Tools. Have CHKDSK look for and fix file system errors.

A new tool to help determine if the drive is really dying is DiskEx or StorEx. This utility is available from Avid Support and ships with new systems. DiskEx or StorEx exercises your disk by making it run through a series of stress tests to see if it will fail completely. If you run one of these utilities and your disk does not fail, then you can confidently search elsewhere for the problem instead of waiting overnight for a new drive, which will not solve your problem.

TECHNIQUES FOR ISOLATING HARDWARE FROM SOFTWARE

Part of the trick of troubleshooting is to determine, before you go very far, whether the problem originates with the hardware or the software. If the problem is internal to the computer hardware, the video and audio cards, then there is very little a novice can do without many cautions and a bit of hands-on training. In early versions of the Media Composer, most systems were Quadra 950-based. The Quadra 950 was very easy to get in and out of and was relatively robust as a chunk of metal and plastic, but it had one drawback: Over time the boards would heat up and cool down and eventually work themselves just a little bit out of the slots. The video and compression boards especially were large and heavy and, if you had the Quadra sitting straight up and down, they would pull themselves out by their own weight. The best strategy if you are still using a Quadra, and hopefully you have upgraded it with a Power PC card, is to lie the computer flat on its side with the side door on the top.

If you really need to get into the Quadra, you should not be stacking drives on it either. If you suspect your board is loose, the only recourse you have is to reseat it. Gently pull the board out of the socket and push it firmly but gently back in. But you must take serious precautions against electrostatic discharge (ESD) and potentially voiding a warranty, so this is not to be done lightly. I'll go into ESD later in this chapter.

If you suspect hardware-related problems, it is always worth a call to Avid Customer Support. If you have determined that the problem is not hardware related, then look to the software. This is the main reason that all Avid Certified Support Representatives are thoroughly trained to understand a wide range of uses of the system. There is nothing more frustrating than having a video engineer or some other technician called to examine the Avid Suite and have them shrug their shoulders and say, "Must be the software," and leave.

SOFTWARE PROBLEMS

There are two major areas where you could have problems in the Avid application: the project files, bins, and User Settings, or the media itself. Try to separate problems with the media playback from problems with the media itself. Media can always be redigitized, but I'll bet you don't have a few extra video cards hanging around. The best way to tell if a particular problem rests with the media is to take a close look at it. Step through the problem areas frame-by-frame. That will show if the problem is there when the drives are not working so hard. Any corrupt images or crazy colors that are visible when you are looking at a still frame can be solved by redigitizing that shot. If you can't see the problem, you must go one step deeper.

Is the problem playback related? See if the sequence plays back without any of the media online. Go to the Finder level and dismount your drives by dragging them to the trash (the drive icons, not the folders inside) or by changing the name of the folder on the media drives from 5.x, 6.x, or OMFI MediaFiles to anything else. I just add an 'x' to the end of the name so I don't get into trouble trying to spell the correct name of the folder again when I am in a hurry. As discussed previously, any change to the name of this very important folder causes all the media inside to go offline. Now does the sequence play? If there are no errors, you know that the problem is related to the drive or the computer board. If you continue to get obscure errors (my favorite is BadMagic), then you need to look to the sequence itself.

There may be something corrupt in your bins, your project, or your User Settings. This is so easy to test that many times it is one of the first things an Avid Support Representative asks you to do. Generally, I suspect this particular problem when something that has been working fine all day stops working or features that should be available are suddenly not there. Create a new bin or project and drag the clips and the sequence you were working on into the new bins.

The next step is to remake your User Settings. This should be a fast check to see if default User Settings clear up the problem. If they do, then spend the time to re-create your settings. Better yet, call up an archived version of your User Settings that has been hidden where no one can get to it. This may also be corrupted, but chances are pretty good that it is not. Again, make sure you have new User Settings for every major change in the software. If you call up a really old User Setting, it may be incompatible with new menus and functions in subtle but important ways.

AUDIO DISTORTION

A common audio problem occurs when you are working along and suddenly the audio starts to sound distorted. If you have heard this audio before, you know it was not digitized at levels that were too high. If you call up the Audio Tool, you can make sure the digitized levels do not go into the red area of the VU meter. It helps to isolate this problem if you have digitized this material yourself and you know that the sound was also recorded correctly in the field. If you have access to the original tapes with the Media Composer, use a button called Find Frame. After pressing the Find Frame button, the system asks you to insert the original tape and cues up to the exact same frame you are viewing in the Editing window.

If the audio levels are good on the tape and have been digitized correctly, but still sound distorted when you play them back in the software, you may have a corrupt file called the DigiSetup. This little file is created automatically, is always there, and sometimes, when it is being updated, gets corrupted. The best thing to do is just track it down and delete it. DigiSetup is in the Macintosh System folder. You can find it with Find File or by opening the System folder and pressing the letter D. Later versions of the editing software make one for you automatically when you launch the application the next time. Otherwise, you need to launch a utility called Pro Tools Setup and just click OK. Another sign that DigiSetup is corrupted is if you can play audio normally for a few seconds, then it just stops or goes silent, yet if you stop and play that section again, there is audio. If the distorted audio is also slightly slowed down, like listening at 95 percent of speed, trashing DigiSetup is always the answer. If all this does not solve your distortion problems, you may need to look to the audio board itself.

FONTS

Even though Apple has tried to make dealing with fonts as simple as possible, somehow it is still complicated. When you try to use the wrong kind of font in the Title Tool, it may look absolutely terrible although all the other fonts look fine. The main complication has to do with what kind of font you have and where it is located. Standard TrueType or Adobe PostScript fonts installed correctly should be available to use in the Title Tool. You should have a copy of Adobe Type Manager (ATM) loaded to deal with the Adobe PostScript fonts.

The confusion comes when people download a font from the Internet and install it or take a single-sized font and try to resize it. These are two different problems. The first situation, a font of unknown origin, generally means that it is a bitmapped font. It is meant to be only one size, and if you try to change it in the Title Tool, it quickly looks chunky with ragged edges. You should really discard this font; it is of no use to you with the Avid.

The second situation is similar, but results from a different cause. When you use Adobe PostScript fonts on a Macintosh, there are two parts to every font—a screen font and a printer font. You may find that one of these, usually the screen font, has not been installed correctly on the system. This means that you no longer have a font that rescales for screen display. Time to go back to the original installation disk and reinstall the font. Sometimes good fonts just go bad. They become corrupt and they must also be reinstalled again. You should always have access to the original software, especially fonts, in case you have to reinstall your Macintosh system software from the CD-ROM.

Incorrect installation may account for fonts not showing up at all. All fonts must go into the Fonts folder in the System folder. You can simplify this by dropping them onto the closed folder icon for the System folder, and the Macintosh asks you if you want them to go into the Fonts folder. You say yes and it does the rest after you restart. The problems occur when you just drag them into the open System folder and drop them inside that open window. Then they are not put away correctly and are loose in the System folder.

Another problem with fonts is that too many of them take up unnecessary amounts of RAM. If you want to slim down your required RAM for the operating system so that you can give the RAM to another application, then removing fonts helps as soon as you restart. Another way to simplify font selection, but complicate your life, is to use font suitcases. Many programs allow you to keep a particular set of fonts together and load them all at once so you have just the ones you need.

MEDIA MANAGEMENT

Another day-to-day concern mentioned before is the management of media on the drives. You may think that you are efficiently deleting all media as you finish each project, but you may be surprised to find all sorts of odd bits of precomputes and imported graphics and so forth floating around on your media drives. The best way to keep a handle on this is to delete media from the Media Tool and not through individual bins. That way you see the project from a big picture point-of-view and can evaluate on a project-by-project basis what must go and what must stay. It also keeps you from accidentally deleting media from another project just because you dragged a duplicate of someone else's master clip into the bin you are now deleting.

The real problem, not only with having many unnecessary small objects on your drives slowing down performance, is that the drives may accidentally become too full. You should always keep a minimum of 50 megabytes on any partition or any drive completely free. There are all sorts of files, like the Mediafile Database, which must occasionally get larger to accommodate the changing nature of the media on the drives. Don't worry about defragmenting media drives, but overfilled drives corrupt media and may crash and take everything with them.

A more difficult problem to diagnose is when you actually have too many files in a single folder. If you are working with nine-gigabyte partitions and digitizing offline resolutions, you may find that on long or complicated projects you are exceeding the limit on the number of files that can be in one folder on the Macintosh. This occurs because Avid must restrict the disk cache to 96k in order to reduce digitizing underruns. This limits the practical number of files in the OMFI MediaFiles folder to around 1200. Above 1200 (or approaching 1200 depending on the size of the files) you may have difficulty booting the system or opening the OMFI MediaFiles folder. If your system is exhibiting these symptoms, you may want to raise the amount of the disk cache, move items to another folder, and change the disk cache back to 96k. If you are always working with offline resolutions, you may want to limit the drive partition size to a maximum of four gigabytes.

SERIAL PORTS

Printers and modems can be connected to your system, but they take up valuable serial port connections. This may be the best reason to use Ethernet for printers since it frees up a serial port so that you can use it for deck control. The modem port and the printer port are identical and interchangeable. It is just that certain software expects to find the modem connected to the modem port and the printer to the printer port (on NT these are just numbered but are still interchangeable). If the software can't find the right device, tell it to look at another port either in the Macintosh Chooser or in the software itself.

Sometimes the serial ports get stuck in an open or closed setting and appear not to recognize that there is a device connected to them. This might happen if you had a printer or a phone connection problem and the last thing you did with that device did not shut it down or turn it off properly. You may notice this in the Avid when you have lost deck control, although everything else is set up correctly. You have the Serial Tool under Tools and you can change which port you are using for deck control or change the port to "None" and change it back to the one you want.

If the Serial Tool doesn't clear the serial port, you must either use a utility called CommCloser that ships with the Avid Macintosh software or zap the PRAM (pronounced pee-RAM). The PRAM is parameter RAM and is a part of the computer memory that holds important information that keeps things like the clock going when the system is shut down. PRAM remembers important settings so that when you boot up, everything is the way you left it. Sometimes you must purge this memory to clear the stuck serial port. TechTool, a piece of freeware from Micromat Computer Systems, is the best utility for this because it zaps the PRAM quickly. Otherwise, you must restart and hold down the P-R-Command-Option keys and wait for the restart chime to sound eight times. This can be pretty tedious if you have a lot of RAM because the restart takes so much longer. You then just restart and reset all the control panels that may have changed back to their default state.

The most important settings to put back are Memory Settings. Modern Memory Manager must be on. With a Quadra, you must turn the 32-bit addressing back on in the Memory Panel. All newer systems should check the Memory Control Panel and turn virtual memory off. If you have a Quadra with a Daystar Power PC upgrade card, you must turn it back on through the Control Panel and shut down completely (not just restart). It can take some time to both go through the procedure and to reset all the control panels, so zapping the PRAM is generally considered a last resort.

OTHER DECK CONTROL TIPS

There are several other reasons why you might not have deck control. When you open the Digitize Tool, you may see the message "No Driver." The first thing to do is to try to reload the deck configuration or force the system to "check decks." Both of these choices are at the bottom of the Digitize Tool under the Deck Model pull-down menu.

If this doesn't work, then you need to do some digging. Obviously, check the cable connections first. You may have a V-LAN or VLX, which is an external deck control device from Videomedia that gives you a wider range of deck control choices. If so, then make sure you are using the V-LAN or VLX cables and not the Avid-supplied deck control cable.

Check your Mac extensions folder to find Serial (Built in) on a PCI system. On an older system, like an 8100, Avid Media Processor, or PowerQuadra, check for the Serial DMA extension.

If none of this works, you may want to check your Release Notes or user manual to make sure you are using a supported deck. Even if the deck is not supported, you should be able to get some limited control using the Generic deck choice in the Deck Template window of the later versions.

WORMS AND OTHER INVERTEBRATES

Lately, the spineless, dirt-eating individuals who like to prove how much damage they can do have been rather busy. There is a common "worm," or self-replicating computer virus, which has infected many Macintosh systems. It is referred to as the Hong Kong virus or Autostart Worm. You may suspect that your system has been infected because it has suddenly slowed in performance and the drives seem to be very busy although nothing is happening. You must try to eliminate this worm before it spreads to other systems that you share files with.

To check to see whether you have been infected, you need to search your drives for invisible files.

- Go to the Find under the File menu.
- Hold down the Option key and click on "name."
- Highlight "visibility."
- Choose "More choices" and search for the letter "D."
- Look for the following files:

      DB
      del DB
      Desktop Print Spooler
      del Desktop Print Spooler

If you find any of these files, then you will need to run an anti-worm program called Wormfood (version 1.3 or later). You can download this program from www.macvirus.com and run it. Do the search again to see if you have been successful (Desktop Print Spooler is a standard Macintosh file, so if that is still there don't worry).

To lessen the chances of becoming infected again you should turn off the Autostart Setting for Quicktime and the CD-ROM.

- Go to the Quicktime Control Panel.
- Uncheck "Enable Audio CD Autoplay" and "Enable CD-ROM Autoplay."

You may want to see if your virus protection software can be updated from the manufacturer's website to protect you in the future. You should probably check this website on a regular basis anyway until the human worms wise up and get a life.

BASIC NT TROUBLESHOOTING

The main objection I hear many editors use to avoid switching from Mac to NT (besides the politics of "monopoly" versus "insanely great") is the fear of troubleshooting a more complicated operating system. The ease of access to the Macintosh System folder is a double-edged sword, and moving vital resources in and out on a whim is the cause for many editing problems. It is also something that many editors have spent time to learn and so can do some basic troubleshooting on their own. Switching to NT means relearning the top ten things to do when a problem arises.

I have used an NT system in my office for normal business needs for over a year and have never had my system crash. I boot it up on Monday morning and shut it down on Friday night. I lock it with a password when I leave my desk and when I leave for the night. Occasionally, a network connection or a printer will have a problem, which is related to that individual server or shared printer, but I have never personally encountered the "blue screen of death" in all that time.

Of course, loading multiple proprietary PCI boards, a terabyte of Fibre Channel storage, SCSI accelerators, ICE boards, and other high-end peripherals does not exactly equate to sending e-mail and making PowerPoint slides. I also do not change the configuration or add new applications very often because I do not have Administrative privileges. This is occasionally frustrating when I need something right away, but I have yet to spend even five minutes troubleshooting this system in the last year.

A serious problem with Avid NT editing systems is when somebody changes something from the original shipping configuration and the system will not boot. This is easily handled by the NT OS during the boot process when it gives you the opportunity to invoke the Last Known Good configuration. Hit the space bar and then type "L" when prompted. This will bypass any changes the last user may have tried and will load a configuration that worked the last time the system booted successfully. You may then try again to get the configuration to be the way you want to be, however, if the present configuration did indeed boot correctly, then it will be the Last Known Good configuration and you will need to look elsewhere to find the problem.

All Avid systems come with three hardware profiles already loaded onto the system. The Avid Configuration (Network Enabled) is the default profile and has all the network information already configured and ready to go. The Avid Configuration (Network Disabled) profile can be chosen during the boot process if you feel that the network is causing your problems (or there is no network connected). It also simplifies the system to make troubleshooting that much easier. And finally, Original Configuration will omit any special Avid hardware drivers except for the screen display driver for the Meridien Display Controller. Again, this would be used for troubleshooting to simplify the system to its basic NT components. This profile must also be chosen during the boot process if you suspect that it is the Avid hardware or hardware drivers that are causing the problem.

Another mode that can be engaged during the boot process is the VGA mode. This would be chosen if you were having problems with the display card or suspected that the Avid display drivers were causing conflicts with something new loaded onto the system. This mode is also extremely useful if you have been messing around with the configurations and suddenly your display becomes unusable. By rebooting and choosing the VGA mode, at least you will have a usable image on your monitor to continue to troubleshoot, reload the Avid display drivers (EDCInstall), or undo what you just did!

STANDARD NT RECOVERY

Every basic NT troubleshooting guide will walk you through two major procedures to make sure you can get your system up and running quickly in the case of problems. These two procedures are creating the NT Setup Diskettes and creating an Emergency Repair Disk

Creating Windows NT Setup Disks

Your system should have arrived with these diskettes already created. You may want to make another set or you may have safely stored these important diskettes in the locked office of your company's accountant (where no one can get them). These are the disks that you will use to boot the system in case of a crash and you are no longer able to boot from the internal C drive.

Use three high-density diskettes and format them. Then load the Windows NT CD-ROM that was used to install the original operating system. Click the Start button and choose Run to bring up the Run dialog box. Know the drive letter of your CD-ROM drive; it is usually F with a standard Avid configuration. If it is F, then type the following text in the Run dialog box:

```
F:\i386\winnt32 /ox
```

Notice the space after winn32 and be sure to substitute your CD-ROM drive letter instead of the F if it is different. Follow the instructions from that point to load the three floppy disks one at a time and label them correctly. Save these floppies in an accessible but safe place. They should be able to get you to a stage that will allow you to either continue troubleshooting or reinstall Windows NT in the case of an emergency.

If you boot the system using Setup disk #1, then you can use the Emergency Repair Disk to repair a problem that might have been caused by a corrupted operating system, user files being deleted, or someone messing with the configuration.

Creating an Emergency Repair Disk

There should have been two Emergency Repair Disks (ERDs) with your Avid system. The first would be the bare bones Factory Default ERD that will help bring your system back to the configuration before any Avid hardware or drivers were installed. This ERD will also help you to troubleshoot a simplified NT system and is an important tool if you need to completely reinstall the NT OS.

The other ERD is the Avid Operational ERD, which is the disk that should be used most often. Generally, there will be something minor that needs to be repaired and you don't need to take the system back to the factory defaults to fix it. This disk (or another disk just like it) should be updated when new applications are added to the system.

To create or update an ERD, you should go back to the Start button and choose Run again. In the Run dialog box type:

```
rdisk /s
```

Notice the space after rdisk and hit Return. This will ask you to insert a high-density floppy, which will be reformatted, and then record the registry information required to reproduce your current setup. Remember that in order for it to return the system to the current configuration, you will need to update the ERD every time you make a change.

Creating a Bootable Disk

If your problems are serious enough, you may just need to boot the system far enough to reinstall the NT operating system from the distribution CD-ROM. A bootable disk has been included with all of the Avid installation software. You can also boot from the three set-up disks discussed earlier.

Reinstalling Avid Software

If you are missing a file or a file has become corrupted, you may want to reinstall only those problem files. By loading the Avid software installation CD-ROM, you can do custom installations to replace drivers, Codecs, or anything else you may suspect as being a problem.

You may want to reinstall the entire Avid application just to start from a clean slate. You may want to move all your AVX plug-ins and AudioSuite plug-ins to another location so they don't get installed over; however, if you suspect that the plug-ins may have something to do with your problem, you may want to reinstall them as well. Be sure you have all the registration information for plug-ins that require being registered the first time you use them.

VERSION NUMBERS

You should always have, either in the back of your head or written down some-place accessible, the version number of almost everything associated with your system. First and foremost is the version number of the current Avid software. You should know it down to the last digit because every small change in the soft-ware has a reason. Avid puts out what they call the "gold" version of the soft-ware, the most tested and most stable version they can achieve within the time allowed before it must be released. Version 6.5 is an example of a gold release. Release dates, as with all software companies, are based on complicated interre-lationships all lining up at the same time. If, after the gold version is shipped, some features didn't make it into the software although they were planned, there may be other releases with an extra decimal point. Release 6.5.2 is an example of that. There may be some procedures that were not tested and appear to have problems, so there may be another revision called a patch release. A patch release is meant to fix one or two small problems. This would be version 6.5.1v2. The important thing to remember about all these releases is that you might not need any of them except the gold version. The other versions have not been tested as thoroughly as the gold release because of the importance of getting out a fix in a timely manner. Most of the time this is not a problem, but asking for a patch release if you don't absolutely need it pushes the envelope unnecessarily.

It is always best to have all the systems in your facility running the same version of the editing software. Most versions are forward compatible, but not backward. This means that a bin created in version 5.5 opens fine in version 6.5, but not the other way around. Once a bin has been opened in the higher version, it has been converted forever to that version. By just opening the bin once in ver-sion 6.5, you may not be able to open it in version 5.5 again. You have to export the bin full of clips as a shot log and import it into the earlier version. Unfortunately, this does not work well for sequences. A bin converter program goes between versions 6.5 and 5.5.1, but it is best not to depend on such compli-cations. If you are forced to change systems in the middle of a job, always insist on the same version of the software or higher. You should know the version of the Macintosh operating system you are using. Some versions are not approved for some models, and you should make sure that your Macintosh can run the lat-est software before you install it. For instance, Apple did not approve the Macintosh Operating System 7.6 for the 9600. Oops. You must use an earlier or later version.

Always make sure you know the latest version of Macintosh operating sys-tem Avid has tested. If at all humanly possible, try to get all your Macs to run the same Macintosh operating software. Keeping everything interchangeable is a valuable goal and should not be complicated by an MIS person or an especially enthusiastic editor who wants to put the newest software on the machine as soon as it is available.

Version numbers also carry over to the hardware. Each of your boards has a revision number, which should be considered during troubleshooting. You may have an old version of a board or a version with a known conflict. You can check the revision of your hardware in several places. If you have Avid software running, look under the Hardware Tool to see each individual board. Also, a utility called Avid System Utility comes on PCI systems, and Slot Looker for NuBus allows you to get more detailed information about each board. This is the better answer if you can't actually launch the Avid editing software. If Avid System Utility or Slot Looker can't see the card at all, calling it an unknown card, or the slot is empty, then that tells you the problem is with the card or the way it is mounted in the slot.

Every utility that ships with your Avid system has a specific version number. The ones that ship are the ones that are meant either for that hardware or that software. Again, grabbing a version from another system just because it is newer may get you into trouble. Certain versions of the Avid Drive Utility, for instance, are designed primarily for four-way drive striping. The rule of thumb is: If the new version has a tangible improvement, completely compatible with your system and supported in the Release Notes, only then should you load it onto your Macintosh.

ELECTROSTATIC DISCHARGE (ESD)

The sneakiest and hardest problem of all to diagnose is one that is very easy to create: damage related to electrostatic discharge (ESD). You may not realize it, but the human body can store and discharge frightening amounts of static electricity. Shuffling across the carpet with a relative humidity of 10 percent generates 35,000 volts! Compare this to the smaller, faster devices that are necessary to achieve the kind of performance necessary to keep the system running happily, and they have a range of susceptibility of several hundred volts.

You can zap a component with static electricity by touching the outside of an ungrounded device or, more probably, by opening the device to do some simple troubleshooting. You may be asked to remove or add RAM or to reseat a troublesome board. Anytime you are going to open a case, be concerned about voiding a warranty or causing ESD damage to the sensitive components inside.

The key to touching anything inside the system is to be grounded. Being grounded ensures that any build-up of static electricity is channeled off to a ground and dissipated. This is best done by wearing a grounding wrist strap and connecting it to a ground or to a metal component inside the computer.

Most important, make sure the computer itself is grounded. The best way to ensure this is to plug the computer into a grounded outlet. An even better solution is to plug the computer into a grounded power strip and then shut the power strip off. That way you are not supplying power to the Mac while you are working on it.

The scariest thing about ESD is that it doesn't always kill—sometimes it just maims. A board or a RAM chip that receives a substantial shock may not fail right away. It may not fail for days, weeks, or months. It may start to show intermittent behavioral problems that cannot be isolated. These are absolutely the worst kinds of problems to troubleshoot because they may not occur for long periods and may not be of a type that points to any one component. It may be the CPU itself that received the shock, and no matter how many boards you replace, it does not solve the problem. This is why ESD should be treated so seriously, any time you handle a component or open the computer, you should be very, very careful.

CALLING CUSTOMER SUPPORT

If, after all these precautions and general maintenance, you must still call Avid Support, at least take heart in a very short, best in class wait time. To make it go even faster, have certain answers prepared since almost all support calls start with the same basic questions. Know your versions, OS, CPU model, and Avid model. Be able to describe what you were trying to do when the problem occurred. This is especially crucial for video engineers who have not taken the time to learn the software. The editor describes the problem and the engineer cannot explain it to the support rep in enough detail. What is the exact wording of the error? Some errors are pretty obscure like "missing a quiesce." Write it down and don't fake it with "something was missing, I think." When did the problem begin? Right after you put in the new RAM? Pull the RAM out. Does the problem happen every time you perform a particular operation or is it really random? Can you repeat it? The simplest answer to any error that seems random is to shut down the Mac and restart.

The other very important thing you need to give to Customer Support is your system ID. If you don't have a system ID, you may not get any support! That is the way Avid determines whether you or your company has a valid support contract. Not something you want to discover at 3 A.M. The system ID can often be found by just doing a Finder search for "sys" or launching the utility Dongle Dumper, but if you can't launch the computer that won't help much. Write it down or use Dongle Dumper and print it out.

We can't deal with all the techniques, potential problems, and error messages in this short space. If you want to know more about your system, there are courses in troubleshooting and a whole curriculum to become an Avid Certified

Support Representative. You don't need to be a technician to keep your system running happily most of the time, but you should have some basic knowledge of what is going on "under the hood." Good maintenance routines and a healthy dose of caution are two necessary items when dealing with sophisticated and complicated systems. Keep it simple and you will be rewarded with fewer steps and less stress when you need to troubleshoot.

14

Nonlinear Video Assistants

With the adoption of new technology has come the blending of post-production responsibilities. Producers and writers become offline editors, offline editors become online editors, and the difference between film and video begins to blur. One thing that this shift has created is many new people who can edit well, but who don't have the inclination, ability, or time to get involved in the technical requirements.

This opens up the possibilities for an entirely new position: the nonlinear assistant. There have always been film assistants whose responsibilities are pretty clear. They handle all the day-to-day requirements of film handling, organization, and preparation for the editor. There have also been video assistants in the past, although their roles have changed through the years and have, at times, been eliminated altogether. It is inconceivable to lack a film assistant on a major feature, but many high-end production companies operate quite well without video assistants.

Assistants are very important if the design of the post-production facility is focused around a central machine room. The editor initiates the communication via an intercom system and tapes are changed, set up, and dubbed by this voice on the other end. Occasionally, that assistant is in the same room and can speed up the editing process by doubling as a sound engineer or a character generator operator.

The elimination or devaluing of the video assistant makes it harder for young people to break into the business. Since the film assistant on a nonlinear project may work a second shift while the editor works throughout the day, there is less opportunity for the interaction between master and apprentice. With tense clients who are paying large sums per hour for time-critical work, the production company that puts an unknown or untested quantity in the driver's seat is taking a risk. They may lose their client forever to the competition or may have to discount the session to appease them.

How then does one break into this business? There are no second video assistants like there would be in film, so what is the entry-level position? Many times it is whatever the facility needs at the time: a dubber, a receptionist, even a courier; however, with the advent of nonlinear editing, the new choice becomes the nonlinear assistant. In large and busy nonlinear post-production facilities, there may be one nonlinear assistant per shift and three shifts per day. The entry level then becomes the graveyard shift and eventually the day or evening shift, where editors can discretely observe skills that keep them in demand. The job doesn't really require the ambition to become an editor, but the people that gain the most from the nonlinear assistant position are those that need to know these subtle skills to move on to the next level.

What responsibilities should such assistants be expected to perform? Much of the knowledge they need has been covered in this book. In fact, many editors perceive many of the techniques in this book as something only an assistant would perform. Others see it as required knowledge before starting a job! It is when a facility desires such a specialization of labor, either for personnel or billable reasons, that the assistants have the most value. They must perform all of the functions and have all the knowledge required to keep the systems running. The post supervisor instructs the assistant on all the requirements to keep as many jobs running smoothly as possible. The administrator or supervisor sees the big picture and the assistant performs the tasks.

These important daily tasks include digitizing, media management, basic maintenance, backup, and output. Anything that is required to prepare the suite for the editor and the smooth transition from one project to another is appropriate for the assistant. Each facility has its own set of requirements, but mastery of all these skills can make someone very valuable to any busy post-production facility.

DIGITIZING

Digitizing also implies following logs, creating bins, setting levels, and understanding drives. The logs are handed off with the understanding that the marked takes or, possibly, everything should be digitized and the master clips named according to the description in the logs. Bad logs mean bad bins unless the assistant knows something about the specific job and is given the freedom to create better master clip names. Each take must be assessed as to whether video or audio only should be digitized, thus maximizing use of disk space. Digitizing way too much by digitizing video for voiceovers or room tone will fill the drives prematurely and cause careful estimates to be off. On the other hand, editors may prefer that, if the drive space exists, everything, even incomplete takes, be kept. Don't assume that just because there isn't enough drive space that certain shots must be left out. It is the assistant's responsibility to find the drives, connect them, and digitize everything if required.

The bins can be named based on tape name, and the editor determines where to put the shots based on content later. Levels, as discussed previously, are crucial to the finishing stage. Distorted audio and blown-out video can come back to haunt the project. All assistants should know how to set levels and monitor them. They may not be perfect levels, but they should be as close as possible to require only a few last-minute tweaks. Learn how to read a vectorscope and waveform monitor or risk being bypassed by the nonlinear finishing revolution!

DRIVES

Understanding drives is crucial to making sure the digitized video can be played back. There is a setting called Drive Filtering under the General Setting. If this is on, only drives capable of playing back the selected resolution are available. Unfortunately, when people use non-Avid drives, they must disable this setting all the time since, when the system does not recognize the firmware loaded onto the media drives, it assumes the drive is incapable of higher resolution playback. This is not always true, but it is safer than assuming every drive can play back every resolution. Not paying attention at this stage can mean having to copy large amounts of material to the proper drives. It is possible to digitize audio to a Zip drive by accident with Drive Filtering turned off! Generally, however, if there is a mismatch between resolution and drive capabilities, there will be an error about either the video or audio overrunning its buffers. Usually, however, this error is because you have set the Macintosh RAM cache too high.

If you can, split the audio and video to separate drives, *not* separate partitions of the same drive. The drives should be named so there is no confusion between what is a new drive and what is just another partition on the same drive. Never fill the drives too full. Leave a minimum of 50 MB free on any partition. Later versions include the ability to split a single media file across multiple drives after it hits a certain length, like over 30 minutes. This will simplify digitizing long clips. They will no longer need to be broken apart by hand to fit on two drives and end up confusing everyone. The problem will lie in separating the two drives with the same clip. Like striped drives, splitting the clip over multiple drives means both drives must be present all the time or the clips are unreadable.

The assistant needs a thorough understanding of SCSI principles. There are many times when fast and wide drives must be combined with a narrow drive. Be sure you can recognize which is which and have the right terminators and cables. Chapter 13 discussed these problems in detail. Unless your facility has installed drives on a very fast network, like Fibre Channel or SSA, you will always be moving drives and checking SCSI IDs. You need to be comfortable with Media Docks or RMAGS. All drives should have unique names so they can be moved to any system and still be identified.

Understanding drives also means knowing how to resuscitate an ailing one and knowing when to call it quits and get another. The number of drives returned to Avid with nothing wrong is astounding. If you can get a drive back to full health in an hour or two, how does that impact the production schedule compared to waiting for the morning rush delivery? What are the replacement policies of your non-Avid drives? Learn how to use all the drive utilities. What is the latest version of the Drive Updater, the program that reloads the drives' firmware, and what are its benefits or drawbacks? Perhaps for your situation the latest Drive Updater would be a disaster. Don't always load the newest thing until you know all the ramifications for your individual needs. AvidDrive Utility or Avid Drive Mounter should be your most important tool here. Use it for mounting, repartitioning, and, as a last resort, a destructive read/write test.

There are important differences between maintenance on a media drive and the internal Macintosh drive. Using the wrong version of Norton's Disk Doctor (only 3.2 and later are approved for striped drives) or SpeedDisk on a media drive can ruin your weekend.

MEDIA MANAGEMENT

All the media on all the drives are under your jurisdiction. Knowing what to keep and what to discard is both incredibly important and commonplace. Media Mover is the most important tool here. It gives each project its own media folder on the media drives. Locking and unlocking media files and deleting precomputes are all regular procedures to be familiar with. Know how long it takes to copy media from one drive to another. Delete through Media Tool or drag entire projects to the trash in their media folders. One of the most common calls to Avid Support concerns the lack of a regular plan to delete precomputes; your knowledge will keep you off the phone and your company humming at the maximum drive capacity.

Understand the network and how it helps you to move media efficiently. Become the network expert if you can, and research the possibilities. Figure out how to improve throughput and how to reduce the number of drives that must be moved.

BASIC MAINTENANCE

If you can, take a Macintosh or Windows NT support class or the Avid Basic Troubleshooting class. If you have some talent at the technical side, consider becoming an Avid Certified Support Representative (ACSR). Always have a floppy that you can boot from and run disk-recovery programs. Be ready to strip all unneeded extensions from a system and know which ones are the absolute minimum to run your specific system. Be ready to do a clean reinstallation of the system software. Between projects do everything you can to make all the systems as similar as possible.

If you aspire to become an editor, there are people who will hold your technical expertise against you. They think you cannot be technically proficient and a true artist. You may have to work a little harder to prove them wrong. It only takes a couple of success stories where you save the system *and* the project before employers see the value in an editor who reduces his or her own downtime.

BACKING UP

Consolidating is the most important feature for backing up. Some people have enough time and tape to back up everything in a project, but more likely, you will be forced to decide what to keep and what to discard. Learn all the variations for consolidating the final sequence and focus your efforts on only the media needed to re-create the sequence.

What happens to a project when it is finished? Create a database that can both retrieve the project and the individual bins. If you print out the bin for each tape and include that with the tape, you have a paper archive when all else fails.

OUTPUT

All forms of output are important and critical to the next step in the project. The Digital Cut may be the master or the approval copy. If it is the master, then the levels must be perfect, and if it is just a VHS, then it must contain all the video tracks of graphics and all the audio, mixed or direct. Do you have a time code generator that can burn in the sequence timecode to the Digital Cut?

The EDL must be rock solid and at least spot-checked for accuracy. A cut list should be scrutinized at every cut because the stakes are so much higher and the chance for adjustment at the next stage more remote. Make as many versions and as many disks as you have time for. Expect that, when the online assembly finally comes, someone may ask for something more. This means that if the request is for video-only EDLs, make a few with audio too just in case. Make several printouts to cover yourself, save time, and help the online editors if there is a mistake at their session. Everybody makes mistakes and, if you have an original copy of the EDL on paper, you can isolate the mistake more quickly as something done wrong after your hand-off.

Talk to professional sound studios. What are the most common mistakes you are likely to make when handing off media and projects to an AudioVision or Pro Tools session? What is right for one house may be wrong for another, maybe because the other facility handles your files wrong! Document all the steps you take to prepare the files and label everything clearly. It really is part of the job to prevent other people from making mistakes.

REDIGITIZING

Does every shot need to be redigitized? It is possible to make an EDL with media offline. If the changes requested pertain only to the open and close of a sequence, then you might be able to get away with digitizing only those sections. You can leave the rest of the media offline.

If the need for redigitizing is because the levels were set wrong, then the saved settings for that tape must be deleted before redigitizing. If some of the footage needs to be digitized in black and white, make sure to set the Compression Tool to monochrome. Even more important, change it back when you're finished! The third monitor shows you a color image all the time because it is monitoring what the signal looks like before it is processed. How many times have people walked into an Avid suite during digitizing at low resolution, looked at the client monitor, and said, "Hey, that doesn't look so bad!" So it does not reflect whether you have left monochrome on or are digitizing at the wrong resolution.

BLACKING TAPE

It is desirable to know everything about connecting and operating video decks. This information is beyond the scope of this book, but it would include knowledge about reference signals, signal termination and loop through, front panel editing, input choices, and blacking tape. After spending an afternoon with a well-known, much-decorated documentary editor, I asked about his deck connections. "What's a BNC?" he said. Although he had a great attitude toward the new technology, there was quite a learning curve involved in making him self-sufficient.

If you are not supported by staff engineers, I highly recommend that you learn how to clean the video heads. Oxide flakes off videotape and sticks in the tiny gap that video heads use to read the video signal. If you have a regular staff of technicians, then *don't touch*, but make sure the video heads are serviced on a regular basis. If no one is regularly cleaning the heads on the video deck, you should take on the responsibility as routine. Cleaning the heads with a proper head-cleaning kit once a week during heavy usage is not a bad idea, and if you are using a consumer deck, of course, there are head-cleaning cassettes. Consumer-quality tape loses oxide faster than professional tape (*The English Patient* was cut using S-VHS and a Film Composer). There is a significant difference between digital and analog cleaning procedures, so make sure you know the deck you are working on. Digital decks do some self-cleaning. Do *not* clean with alcohol. Alcohol leaves a gummy film when it dries. Use a freon substitute and a wipe that is recommended by the manufacturer. Many people like TekWipes. Some rental facilities cover the edges of the VTR top cover with a seal so they know if it was opened and tampered with. Check with the rental company before breaking the seal.

For connecting Betacam SP decks, or any decks that have both composite and component input, I recommend looping the blackburst signal out of the reference input to the composite input. This serves several purposes, but the most important is that you will never accidentally use the composite input for your master tape digital cut. It is always a black signal. The other benefit is that you can change the position of the switch on the front panel of the deck that changes the deck input from component to composite. Then you can begin to black tape without disconnecting or reconnecting any cables. Always monitor the output of any deck that is recording.

If you buy tapes by the case, then a good practice is to black them all whenever there is any downtime. By turning down the audio inputs and switching over to composite video input, which has been connected to a source of black, you can black tapes at a moment's notice throughout the day. Be sure to turn the audio inputs back up (or pop them back into the preset position) and throw the front panel switch back to YRB (component) before you start a digital cut!

When blacking tapes, set the four switches under the front panel of the Betacam deck to:

- DF (drop-frame) or NDF (non-drop-frame) if in NTSC
- Internal
- Record run
- Preset

This ensures that the timecode is generated from the internal timecode generator. It is not looking for an external signal. The timecode only increments (runs) when the tape is recording, and the starting point is wherever you preset it to start, no matter what other timecode is on the tape.

The ability to preset a starting number for timecode is achieved by manipulating the buttons on the outside of the front panel. Each deck is slightly different, but on Sony decks:

- Press the Hold button.
- Use either the "+" and "−" buttons or the search knob to choose both hours, minutes, seconds, and frames and what those numbers will be.
- Press the Set button.

Although the number in the LED does not show it yet, when you begin recording, the timecode starts incrementing from the desired preset number.

Other switches to check on a Betacam SP deck are on top of the front panel (as opposed to on its face or hidden by it), where the record inhibit switch is also located. The 2/4 field switch can sometimes be the cause of field-based problems when digitizing graphics from tape. You should always digitize everything with this switch in the four-field position (NTSC).

If you are trying to lay four discreet channels of audio to the deck, you'll need to switch the front panel audio switch that says 1/2 and 3/4. If it's on 1/2, then the audio sent to 1 and 2 is automatically laid to 3 and 4 if those input levels are up. Many people use channels 3 and 4 on the Betacam SP because of the superior signal-to-noise ratio. They may want four discreet channels from the Avid to the deck so the stereo music can go to channels 3 and 4 while the narration and sound bites go to channels 1 and 2. A final note of caution: Tracks 3 and 4 are recorded helically, alongside the video. If you make a video insert edit on a Betacam SP master, you will wipe tracks 3 and 4 for the length of the edit!

There may be more specific requirements for assistants at your facility, but these should cover the basic skills to make you useful from day one. All markets are slightly different with different terminology and different kinds of clientele. Every chance you get to watch the editors work should be snatched up, not just to see how they use the interface, but also what they are doing with it. Observe the way creative ideas are thrown around, accepted, rejected, modified, and experimented. This is always more important than the equipment being used.

You have an advantage over assistants in the past because, if you can get permission, you can take the opportunity to cut your own versions of the scenes, commercials, or segments when the suite would otherwise be empty. You can't mess up the film or add wear to the tapes just dissecting what the editor did and trying a few variations of your own. If you are lucky, you can get the editor to check them out and offer suggestions. A recommendation by the editor may move your career along faster than anything else.

Positions change and so does technology. If you stay flexible, always willing to learn something new, you will always stay valuable to your employer. This book outlined some of the more important techniques to keep in mind with the Avid editing systems, but by no means all of them. To keep up with the speed of change takes more effort and more research than you might have expected. Stay focused on the most important aspect of the technology — the storytelling — and you will never be out of date.

Index

ABOUT THE AUTHOR

Steve Bayes has been designing and organizing editor workshops and seminars around the world since 1993. He has recently served as Senior Instructor at Avid Technology, Inc, where he has assisted in the development of the courseware being used by the Avid Authorized Education Centers and by the 100 Avid Certified Instructors for Media Composer and MCX. As an editor, postproducer and consultant Steve has worked extensively in advertising, educational and long format television and video, and has been the recipient of numerous awards. He is currently a designer of the Avid Media Composer.